EVOLUTION
THE HUMAN STORY

EVOLUTION
THE HUMAN STORY

REVISED EDITION

DORLING KINDERSLEY

Senior Editor Peter Frances
Senior Art Editor Ina Stradins
Project Editor David Summers
Managing Editor Angeles Gavira Guerrero
Managing Art Editor Michael Duffy
US Editor Kayla Dugger
Jacket Editor Claire Gell
Jacket Designer Surabhi Wadhwa
Jacket Design Development Manager
Sophia MTT
Pre-production Producer Andy Hilliard
Producer Alex Bell
Associate Publisher Liz Wheeler
Publishing Director Jonathan Metcalf
Art Director Karen Self
Design Director Phil Ormerod

FIRST EDITION

Senior Editor Angeles Gavira Guerrero
Senior Art Editor Ina Stradins
Project Editors Gill Pitts, Steve Setford, David Summers, Miezan van Zyl
Designers Dave Ball, Paul Drislane, Duncan Turner, Steve Woosnam-Savage
Assistant Designers Riccie Janus, Fiona Macdonald
Junior Design Assistant Jonny Burrows
Jacket Designer Mark Cavanagh
Production Editor Ben Marcus
Production Controller Erika Pepe
Picture Researcher Liz Moore
Photographer Gary Ombler
Scenic illustrations Robert Nicholls
Managing Editor Sarah Larter
Managing Art Editor Michelle Baxter
Art Director Phil Ormerod
Associate Publisher Liz Wheeler
Publisher Jonathan Metcalf

3D model reconstructions by Kennis & Kennis

This American Edition, 2018
First American Edition, 2011
Published in the United States by DK Publishing
345 Hudson Street, New York, New York 10014

Copyright © 2011, 2018 Dorling Kindersley Limited
Foreword copyright © Dr Alice May Roberts, 2011
DK, a Division of Penguin Random House LLC
18 19 20 21 22 10 9 8 7 6 5 4 3 2 1
001–306253–July/2018

I rights reserved. Without limiting the rights under the copyright reserved above, no part of this
ublication may be reproduced, stored in or introduced into a retrieval system, or transmitted, in
y form, or by any means (electronic, mechanical, photocopying, recording, or otherwise), without
the prior written permission of the copyright owner.
Published in Great Britain by Dorling Kindersley Limited.

A catalog record for this book is available from the Library of Congress.

ISBN: 978-1-4654-7401-8

DK books are available at special discounts when purchased in bulk for sales promotions,
remiums, fund-raising, or educational use. For details, contact: DK Publishing Special Markets,
345 Hudson Street, New York, New York 10014
SpecialSales@dk.com

Printed and bound in China

A WORLD OF IDEAS:
SEE ALL THERE IS TO KNOW
www.dk.com

CONTENTS

174 OUT OF AFRICA

198 FROM HUNTERS TO FARMERS

AUTHORS AND CONSULTANTS

Understanding Our Past

Author: **Professor Michael J. Benton**, Professor of Vertebrate Palaeontology, University of Bristol, UK.
Consultants: **Dr. Fiona Coward** (see Hominins); **Professor Paul O'Higgins**, Professor of Anatomy, Centre for Anatomical and Human Sciences, Hull York Medical School, UK.

Primates

Author: **Professor Colin Groves**, School of Archaeology and Anthropology, Australian National University, Canberra, Australia.
Consultant: **Professor Eric J. Sargis**, Department of Anthropology, Yale University / Curator of Vertebrate Paleontology, Peabody Museum of Natural History, USA.

Hominins

Author: **Dr. Kate Robson-Brown**, Senior Lecturer in Biological Anthropology, Department of Archaeology and Anthropology, University of Bristol, UK, with contributions by **Dr. Fiona Coward**, Senior Lecturer in Archaeology and Anthropology, Bournemouth University, UK.
Consultant: **Professor Katerina Harvati**, Head of Paleoanthropology, Institute of Early Prehistory and Medieval Archaeology and Senckenberg Center for Human Evolution and Paleoecology, Eberhard Karls Universität Tübingen, Germany.

Out of Africa

Author: **Professor Alice Roberts**, Professor of Public Engagement in Science, University of Birmingham, UK. Consultant: **Dr. Stephen Oppenheimer**, School of Anthropology, University of Oxford, UK.

From Hunters to Farmers

Author: **Dr. Jane McIntosh**, Senior Research Associate, Faculty of Asian and Middle Eastern Studies, University of Cambridge, UK.
Consultant: **Dr. Peter Bogucki**, Associate Dean for Undergraduate Affairs, School of Engineering and Applied Science, Princeton University, USA.

Revised Edition

Revisions by **Dr. Fiona Coward**

FOREWORD

As humans, we are aware of ourselves. We each have a strong sense of self that emanates from our unique consciousness, and which seems to naturally lead us to ask questions about who we are, and where we come from. Like no other animal, we seem to have a very deep-seated need to know ourselves.

For thousands of years, humans have attempted to answer questions about our origin, our place in the natural world, and our relationship with other forms of life. Religion and philosophy may provide one way of exploring these questions, but science leads us to look for evidence and answers in the world around us, and within us. This empirical approach to age-old questions has revealed extraordinary secrets from our past, allowing us to reach far back in time to investigate our family tree and to meet long-dead ancestors. An evolutionary perspective offers us a deep and rich understanding of ourselves, and places us, as a species, in our own biological and ecological context.

We are primates, and this book starts by introducing our living relatives in this group. The next chapter takes us back to the roots of the human family tree, and we meet our ancestors. The Kennis brothers—whose artful reconstructions I have long admired—have produced a host of extraordinary,

RECORD OF OUR PAST
The rocks around Lake Turkana in Africa's Great Rift Valley contain the remains of several species of human ancestors from up to 4 million years ago. Some of these fossils have played a pivotal part in shaping our ideas about human evolution.

This is not a story of an inexorable rise to power, to worldwide domination and dominion over the rest of the natural world. It is not a story of an inevitable and linear progression, from a life in the trees to great civilizations. Evolution through natural selection may tend to produce greater diversity and complexity over time, but that is not the same thing as "progress." Evolution unfolds in unpredictable and surprising ways, and it is both humbling and wonderful to realize that there was nothing inevitable about the appearance of our own species on earth; serendipity underlies the greatest achievements of our civilizations.

ALICE ROBERTS

UNDERSTANDING OUR PAST

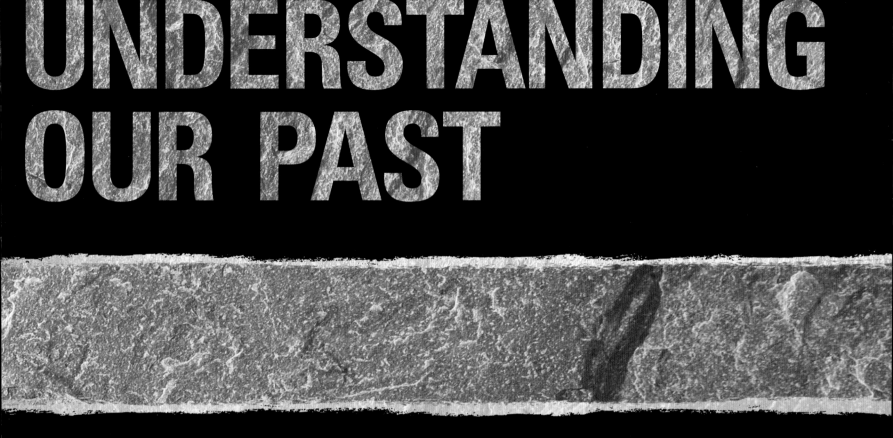

The roots of modern humans can be traced back into deep time, many millions of years ago. The primary evidence comes from fossils—skeletons, skulls, and bone fragments. Scientists have many tools that allow them to extract subtle information from these ancient bones. It's even possible to extract ancient DNA from some specimens. The environmental contexts in which fossils are found, as well as artifacts made by some of our more recent ancestors, also provide useful clues. Modern scientific research in the field and in the laboratory can now provide a rich understanding of how our ancestors lived.

Back in time

Deep time, during which the Earth was formed, extends many millions of years into the past. Its study is the reserve of geology, the study of the Earth. Geologists now know a huge amount about how the surface of the Earth has changed over millions of years and the rocks that make up the Earth's surface can be dated with increasing accuracy.

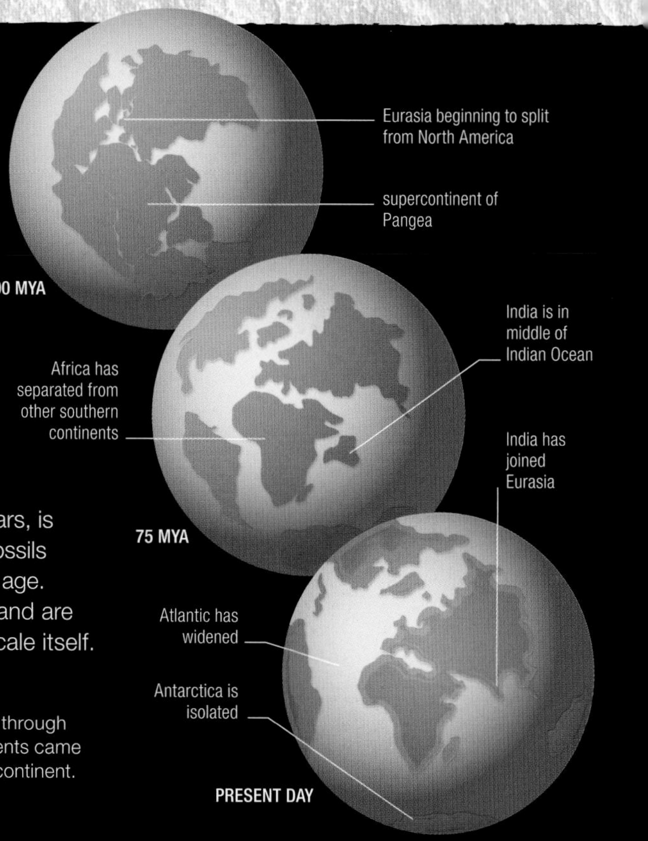

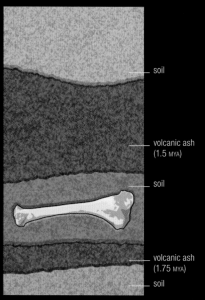

200 MYA
- Eurasia beginning to split from North America
- supercontinent of Pangea

75 MYA
- Africa has separated from other southern continents
- India is in middle of Indian Ocean
- India has joined Eurasia

PRESENT DAY
- Atlantic has widened
- Antarctica is isolated

Geological time

The Earth formed about 4,500 million years ago (MYA), from condensing material in space. At first, the whole planet was molten, and there was no solid crust, water, or atmosphere. After 500 million years, the surface had cooled to form rocks and later, as the surface cooled further, oceans accumulated. The first simple microbial life emerged about 3,600 MYA. Over the next 3,000 million years, life remained relatively simple, progressing from microscopic to visible forms—this whole time is termed by geologists the Precambrian. The remainder of geological time, the past 544 million years, is known in much greater detail because fossils first became widespread in rocks of that age. Fossils tell us about the life of the time, and are used to determine the geological timescale itself.

SHAPE SHIFTING
These globes show the dance of the continents through time. Two hundred million years ago, the continents came together, with Africa and India part of one supercontinent. The various continents have since split apart.

Precambrian

- 4,500 MYA Formation of Earth
- 3,600 MYA First evidence of life
- 2,400 MYA First evidence of bacteria
- 1,850 MYA First eukaryote

4,500	4,000	3,500	3,000	2,500	2,000	1,500	1,000

MILLION YEARS AGO (MYA)

THE DIVISION OF GEOLOGICAL TIME
The geological timescale is divided into eras and periods that were named largely in the 1830s and 1840s. This standard scale is under constant revision and improvement.

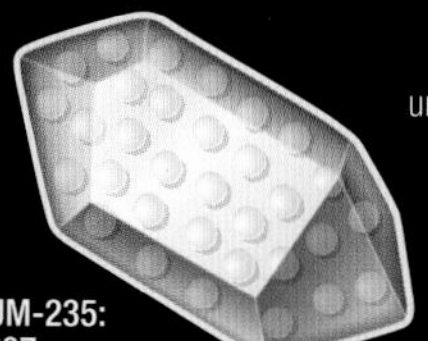

Paleozoic

Cambrian	Ordovician	Silurian	Devonian
535 MYA First fish		425 MYA Oldest land plant	
	485 MYA First vertebrates with bone		380 MYA First spiders
450 MYA First evidence of land arthropods		375 MYA First amphibian, first colecanth	
544 MYA	488.3 MYA	443.7 MYA	416 MYA

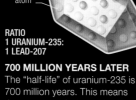

uranium-235 atom

RATIO
1 URANIUM-235: 0 LEAD-207

YEAR OF FORMATION
A mineral crystallizes from molten rock and contains radioactive uranium-235 atoms (shown here in yellow). Uranium-235 will eventually decay to form the isotope lead-207.

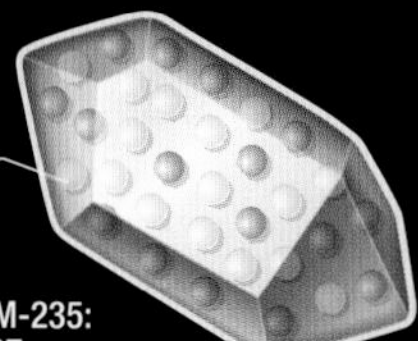

RATIO
1 URANIUM-235: 1 LEAD-207

700 MILLION YEARS LATER
The "half-life" of uranium-235 is 700 million years. This means that over a period of 700 million years, 50 percent (half) of the uranium atoms will decay to become lead-207 isotopes.

lead-207 atom

RATIO
1 URANIUM-235: 3 LEAD-207

1,400 MILLION YEARS LATER
When another 50 percent of the remaining uranium-235 atoms has decayed, the ratio of uranium to lead atoms is 1:3. This means that 1,400 million years has passed since the formation of the rock.

RATIO
1 URANIUM-235: 7 LEAD-207

2,100 MILLION YEARS LATER
If a geologist measures the ratio remaining in the rock (now 1:7), the rock is dated to three half-lives. Because the half-life of uranium-235 is 700 million years, the rock is 2,100 million years old

Dating the timescale

The standard geological timescale is based on regional-scale studies of geology, fossils, correlation, and absolute dating. Regional geological work began over 200 years ago, when geologists in England, France, and Germany noticed that rock layers occurred in predictable sequences, and these could often be identified by specific fossils they contained (see p.15); same fossils, same age of rock. Thousands of correlations of fossils from place to place produce a detailed relative timescale, in millions of years.

Absolute dating

Exact geological dates come from radioactive minerals in particular rocks, especially volcanic ashes within sequences of sediments. When the molten ash settled, the structures of many newly formed minerals were fixed. Radioactive minerals such as uranium (see left) or potassium decayed at a predicable rate, over millions of years. The proportions of the original ("parent") mineral to the decay ("daughter") product give a measure of the age of the rock when calibrated against the known rate of decay. Carbon–14 dating is another form of radiometric dating that works on the principle of measuring decay, but it only works on organic material up to 50,000 years old

- soil
- volcanic ash (1.5 MYA)
- soil
- volcanic ash (1.75 MYA)
- soil

DATING FOSSILS
Radiometric dating of fossils depends upon dating the nearest appropriate igneous rocks, such as volcanic lava and ash deposited as part of the strata. Dating igneous rock above and below a fossil gives the maximum and minimum possible age

Tectonics and life

The idea of continental drift was proposed by the German meteorologist Alfred Wegener (1880–1930) in 1912, based on two lines of evidence. First, he noted the close match of the coastlines on the Atlantic Ocean, and how, for example, the east coast of South America "fitted" the west coast of Africa. His suggestion that the Atlantic had once not existed, and that all continents had been joined as one great supercontinent, termed Pangea (Ancient Greek for "all world"), was confirmed by his observation of shared rocks and fossils across southern lands dating to the Permian and Triassic, some 250 million years ago.

At first, geologists opposed the idea of continental drift because there did not seem to be any proof for how continents could move. But evidence built in the 1950s and 1960s showed how slow movement of molten magma beneath the crust drives the process.

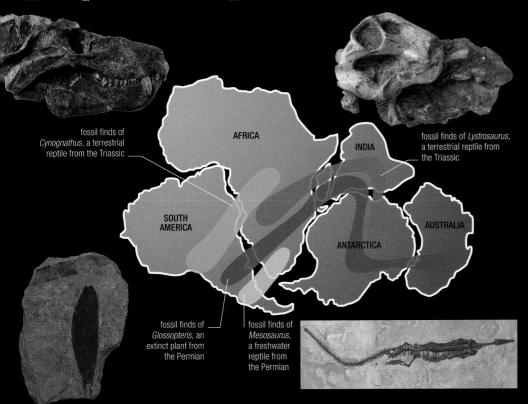

fossil finds of *Cynognathus*, a terrestrial reptile from the Triassic

AFRICA

INDIA

fossil finds of *Lystrosaurus*, a terrestrial reptile from the Triassic

SOUTH AMERICA

ANTARCTICA

AUSTRALIA

fossil finds of *Glossopteris*, an extinct plant from the Permian

fossil finds of *Mesosaurus*, a freshwater reptile from the Permian

LIFE ON THE CHANGING PLATES
Fossils of the same species have been recovered on different continents, proving that these areas were once joined. The reptile *Mesosaurus* was found in a part of Africa and also in South America, which join perfectly when the continents are brought together.

Mesozoic

Carboniferous	Permian	Triassic	Jurassic	Cretaceous
320 MYA First conifers		215 MYA First mammals		125 MYA First flowering plants
		250 MYA Mass extinction		155 MYA First bird
		225 MYA First dinosaurs		65 MYA Extinction of dinosaurs
359.2 MYA	299 MYA	250 MYA	199.6 MYA	145.5 MYA 65 MYA

Changing climates

Written historical records reveal major changes in temperatures over tens or hundreds of years, as people often recorded extreme weather when the normal crop cycles were disturbed. By studying the chemical composition of ice cores drilled from ice sheets and glaciers we can look back to 400,000 years ago or more at detailed temperature fluctuations on a yearly, decade, and century cycle. These records show glacial and interglacial cycles, as the North Pole ice cap expanded and reduced. Geologists can also document changes in climate over millions of years by studying sediments on lake and ocean beds, fossilized corals, and tree rings of fossilized trees.

THE LAST ICE AGE
An extreme example of climate change is the Quaternary glacial phases. The last glacial was marked by the expansion of the polar ice cap over Siberia, Europe, and Canada. Smaller ice caps expanded outward from the Rockies, the Alps, and the Himalayas.

CARBON DIOXIDE AND TEMPERATURE
The carbon dioxide curve (above), documented from an Antarctic ice core, shows four major cycles over the past 400,000 years, with peaks corresponding to peaks in temperature (interglacials), and troughs marking the cold glacial phases. The last glacial ended 12,000 years ago.

temperature (blue)

carbon dioxide (purple)

CARBON DIOXIDE (PARTS PER MILLION BY VOLUME)

300
280
260
240
220
200
180

TEMPERATURE CHANGE °C (°F)

4 (7.2)
2 (3.6)
0
-2 (-3.6)
-4 (-7.2)
-6 (-10.8)
-8 (-14.4)
-10 (-18)

400,000 350,000 300,000 250,000 200,000 150,000 100,000 50,000 0
AGE (YEARS BEFORE PRESENT)

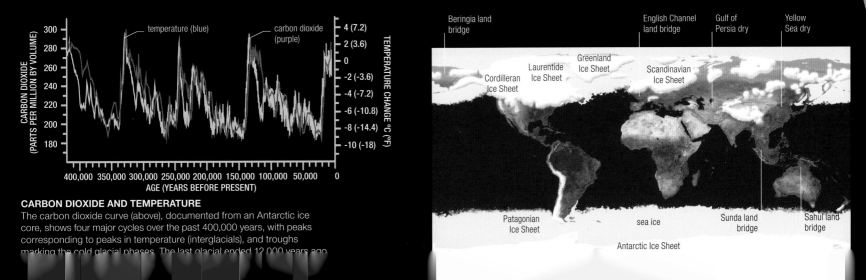

Beringia land bridge

English Channel land bridge

Gulf of Persia dry

Yellow Sea dry

Greenland Ice Sheet

Laurentide Ice Sheet

Scandinavian Ice Sheet

Cordilleran Ice Sheet

Patagonian Ice Sheet

sea ice

Sunda land bridge

Sahul land bridge

Antarctic Ice Sheet

Parade of life

The Cenozoic Era began with the catastrophic extinction (referred to as the K–T Extinction) of the dinosaurs, probably triggered by a major meteorite impact on what is now southern Mexico. This time was also marked by much wider extinctions in the sea and on land, with many plant and animal groups dying out or suffering substantial reductions. The Cenozoic began with a somewhat depleted world, full of opportunities for animals and plants that could evolve to fill the gaps left by the dinosaurs. For the first 20 million years of the Cenozoic, it was not clear that mammals would be the main beneficiaries—during this time, the role of top carnivore was taken by giant birds or crocodiles in different parts of the world. Huge flightless birds fed on horse ancestors in Europe and North America, and on other dog-sized mammals in South America. The ancestors of modern cats and dogs emerged later, and they eventually rose to dominate.

Spread of the grasslands

Climates were famously warm during the age of the dinosaurs, and temperatures continued to decline from about 100 million years ago. The Cenozoic Period was predominantly a time of cooling climates. As polar climates cooled, the centers of the continents became drier. In some cases, this gave rise to deserts, but in others to wide expanses of grasslands. Grasses had been minor elements of the vegetation, but they became hugely important, as they are today, with the evolution of prairies 20–25 million years ago on all continents.

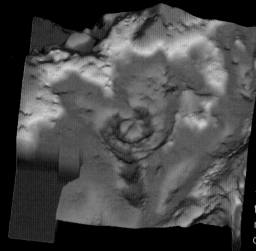

CHICXULUB CRATER
This image is a 3-D geophysical reconstruction of the buried crater, lying partly onshore and partly offshore, in southern Mexico. The crater is over 93 miles (150 km) in diameter, and indicates the impact of an asteroid about 6 miles (10 km) in diameter.

REDWOOD RELATIVE
Metasequoia occidentalis was abundant in the American Midwest in the Cenozoic, when climates were warmer and wetter. The species has died out in the USA because of modern arid conditions.

paired leaves on opposite sides of stem

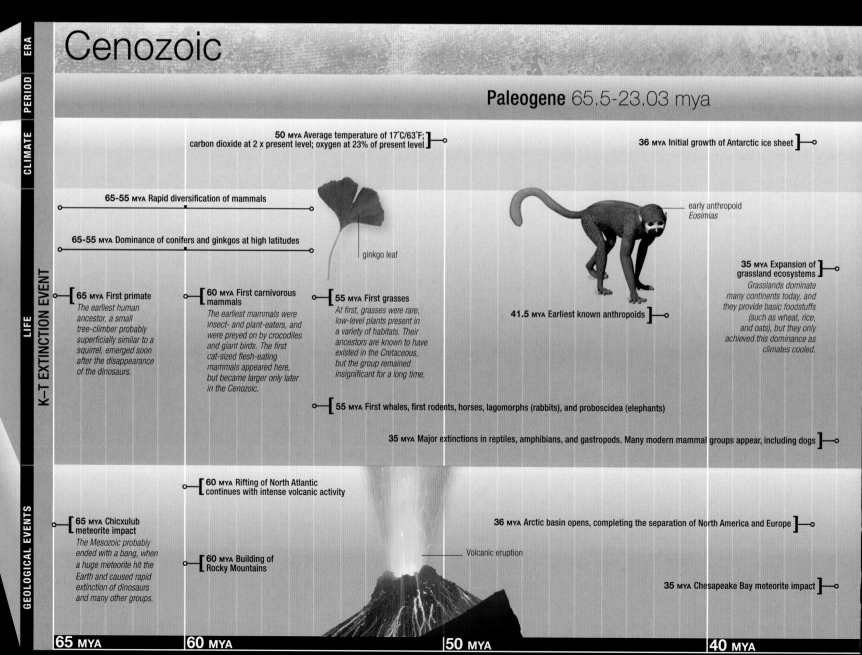

Cenozoic

ERA | PERIOD | CLIMATE | LIFE | GEOLOGICAL EVENTS

Paleogene 65.5-23.03 mya

50 MYA Average temperature of 17°C/63°F; carbon dioxide at 2 x present level; oxygen at 23% of present level

36 MYA Initial growth of Antarctic ice sheet

65-55 MYA Rapid diversification of mammals

65-55 MYA Dominance of conifers and ginkgos at high latitudes

ginkgo leaf

early anthropoid *Eosimias*

35 MYA Expansion of grassland ecosystems
Grasslands dominate many continents today, and they provide basic foodstuffs (such as wheat, rice, and oats), but they only achieved this dominance as climates cooled.

K–T EXTINCTION EVENT

65 MYA First primate
The earliest human ancestor, a small tree-climber probably superficially similar to a squirrel, emerged soon after the disappearance of the dinosaurs.

60 MYA First carnivorous mammals
The earliest mammals were insect- and plant-eaters, and were preyed on by crocodiles and giant birds. The first cat-sized flesh-eating mammals appeared here, but became larger only later in the Cenozoic.

55 MYA First grasses
At first, grasses were rare, low-level plants present in a variety of habitats. Their ancestors are known to have existed in the Cretaceous, but the group remained insignificant for a long time.

41.5 MYA Earliest known anthropoids

55 MYA First whales, first rodents, horses, lagomorphs (rabbits), and proboscidea (elephants)

35 MYA Major extinctions in reptiles, amphibians, and gastropods. Many modern mammal groups appear, including dogs

60 MYA Rifting of North Atlantic continues with intense volcanic activity

65 MYA Chicxulub meteorite impact
The Mesozoic probably ended with a bang, when a huge meteorite hit the Earth and caused rapid extinction of dinosaurs and many other groups.

60 MYA Building of Rocky Mountains

Volcanic eruption

36 MYA Arctic basin opens, completing the separation of North America and Europe

35 MYA Chesapeake Bay meteorite impact

65 MYA | **60 MYA** | **50 MYA** | **40 MYA**

Polar ice caps emerged later, some 15 million years ago. Ice caps are self-generating once they are big enough: at first, the small winter ice cover melts in summer, but once the sheet is wide enough, it reflects sunlight, keeping its temperature lower than the surrounding air (called the albedo effect), thereby preventing the ice from melting.

tall spines in ridge
along back

LAST DINOSAUR
Dinosaurs such as this plant-eating *Corythosaurus* existed to the very end of the Cretaceous period, and then disappeared rather suddenly.

Glacial periods

The Quaternary, sometimes called the "great ice age," spans the last 2.6 million years. It witnessed many glacial episodes of varying intensity. Northern continents were covered by thick ice sheets, affecting climates worldwide. When the ice advanced, plants and animals kept pace, generally moving south. In interglacial episodes, warm Mediterranean and Caribbean climates extended as far north as what is today London and New York.

Northern mammals

Side by side with early humans in Europe, most notably the cold-adapted Neanderthals, lived other glacial animals such as wooly rhinos, mammoths, cave bears, cave lions, reindeer, and Arctic foxes. These animals were able to feed on sparse winter foods such as lichens, and they gorged on a wider array of plants that blossomed during the short summers. As the ice retreated, the reindeer and Arctic foxes moved north with the ice, but the larger mammals died out as their habitat areas shrank.

GLACIER
The head of the Perito Moreno glacier in Argentina shows the massive amounts of water locked up in large ice sheets, and the great transformative power of glaciers to shape the landscape.

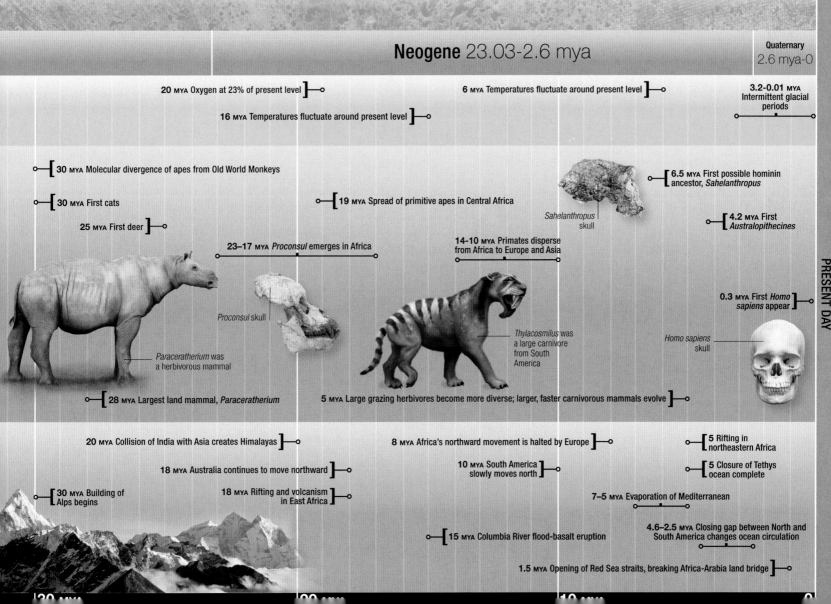

Neogene 23.03-2.6 mya

Quaternary
2.6 mya-0

20 MYA Oxygen at 23% of present level

6 MYA Temperatures fluctuate around present level

3.2-0.01 MYA
Intermittent glacial periods

16 MYA Temperatures fluctuate around present level

30 MYA Molecular divergence of apes from Old World Monkeys

30 MYA First cats

25 MYA First deer

19 MYA Spread of primitive apes in Central Africa

23–17 MYA *Proconsul* emerges in Africa

14-10 MYA Primates disperse from Africa to Europe and Asia

Sahelanthropus skull

6.5 MYA First possible hominin ancestor, *Sahelanthropus*

4.2 MYA First *Australopithecines*

Proconsul skull

Thylacosmilus was a large carnivore from South America

Paraceratherium was a herbivorous mammal

0.3 MYA First *Homo sapiens* appear

Homo sapiens skull

PRESENT DAY

28 MYA Largest land mammal, *Paraceratherium*

5 MYA Large grazing herbivores become more diverse; larger, faster carnivorous mammals evolve

20 MYA Collision of India with Asia creates Himalayas

8 MYA Africa's northward movement is halted by Europe

5 Rifting in northeastern Africa

18 MYA Australia continues to move northward

10 MYA South America slowly moves north

5 Closure of Tethys ocean complete

30 MYA Building of Alps begins

18 MYA Rifting and volcanism in East Africa

7–5 MYA Evaporation of Mediterranean

4.6–2.5 MYA Closing gap between North and South America changes ocean circulation

15 MYA Columbia River flood-basalt eruption

1.5 MYA Opening of Red Sea straits, breaking Africa-Arabia land bridge

The geologic record

Early observers thought that the rocks and minerals under their feet were randomly arranged, and could be classified only as useful (such as gold, coal, and building stone) or useless. However, several early geologists realized that there was a predictable pattern, and that the rocks actually provided key evidence of the Earth's history.

Layers of rock

All rocks started as molten magma. Igneous rocks form when magma cools and solidifies. Tectonic plate movements push up great mountain ranges, which are immediately subject to erosion by wind and water. The eroded grains are carried downstream and deposited in rivers, lakes, and seas. The rocks formed from these accumulated sediments are called sedimentary rocks. Extreme heat and pressure can transform existing igneous and sedimentary rocks, bending and baking them to form metamorphic rocks.

Earth's history is read from layered rocks, or strata. Sedimentary rocks provide much of this information, although layers of lava and ash may appear between sediments, and metamorphic rocks can also prove useful. The structure of these layers indicate the sequence of events in the Earth's past.

fine-grained matrix indicates rapid cooling

IGNEOUS ROCKS
Crystalline rocks, such as granite or basalt, were once molten, and cooled either slowly or fast, forming either large or small crystals.

evidence of bedding

SEDIMENTARY ROCKS
Layers of rocks such as limestone or sandstone formed from accumulated sediments, and may be tens of miles thick.

biotite

METAMORPHIC ROCKS
Metamorphic rocks may show evidence of the great forces and temperatures that formed them: the biotite in this schist shows a medium temperature of formation.

WILLIAM **SMITH**

The English geologist William Smith (1769–1839) was the first to set the geologic record in order, working out stratigraphic sequences and drawing geologic maps. Smith was a practical geologist who earned a living plotting routes for canals. To do this he mapped the rocks, and noticed how the same fossil assemblages occurred in the same order wherever he looked.

The fossil record

Key to our understanding of Earth history is the fossil record: the mineralized remains of ancient plants and animals. William Smith (see panel, opposite) identified the core principles of biostratigraphy, which is the science of ordering rock layers (or strata) and correlating (matching) them from place to place based on the fossils they contain. The key elements of the geologic timescale (see pp.10–11) were completed by 1840, and this scheme, first drawn up in Europe, was then applied across other continents. Fossils occur in the same time ranges in each part of the world, and some, called index fossils (see right), are used as reference points to distinguish well-defined intervals of time. The fossil-based timescale has been applied worldwide, and ever-finer details are worked out.

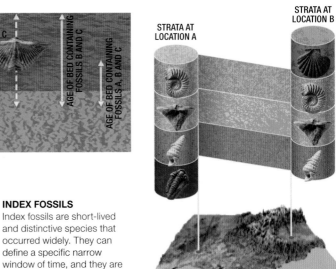

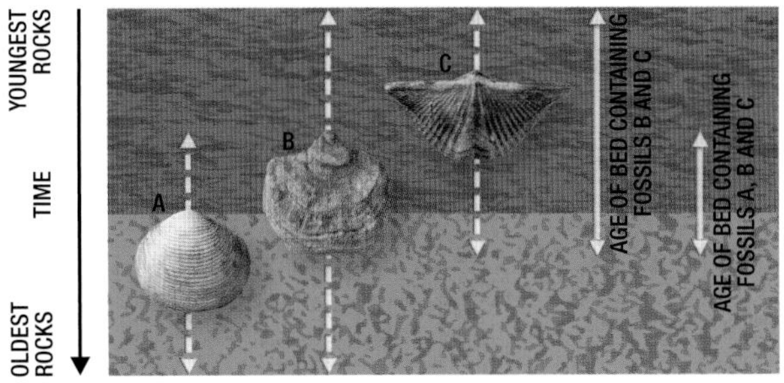

YOUNGEST ROCKS

TIME

OLDEST ROCKS

AGE OF BED CONTAINING FOSSILS B AND C

AGE OF BED CONTAINING FOSSILS A, B AND C

STRATA AT LOCATION A

STRATA AT LOCATION B

FOSSIL ASSEMBLAGES
Each fossil species had a definite range in time, from its origin to its extinction. Careful recording of ranges and how they overlap allow paleontologists to identify particular timelines in many parts of the world.

INDEX FOSSILS
Index fossils are short-lived and distinctive species that occurred widely. They can define a specific narrow window of time, and they are used to match strata of the same age from place to place.

GEOLOGIC SUCCESSION
The buildup of rocks over millions of years can develop a "layer-cake" stratigraphy, just as William Smith saw in the south of England. Each geologic period has a characteristic fossil found in the sedimentary rock that formed during that period. An example fossil for each period from the Triassic onward is shown here.

251 MYA TRIASSIC PERIOD
The Triassic in England was a time of lakes and rivers. Red mudstones and sandstones accumulated, some containing *Dicellopyge*, a freshwater fish from this period.

199 MYA JURASSIC PERIOD
The land was flooded by warm, shallow seas in the Jurassic, and fossils such as the marine reptile *Ichthyosaurus* are commonly found.

145 MYA CRETACEOUS PERIOD
Terrestrial and marine sediments accumulated in the Cretaceous. The shells of sea urchins such as *Micraster* are common fossil in Cretaceous chalk beds.

65 MYA TERTIARY PERIOD
Sea levels fell and climates cooled in the Tertiary. Fossils of the freshwater stingray *Heliobatis* are found in lake sediments from this time.

2.6 MYA QUATERNARY PERIOD
Seas advanced and retreated as the ice sheets shrank and grew. Human fossils are found at all phases of these climatic cycles, often near coasts or lakes.

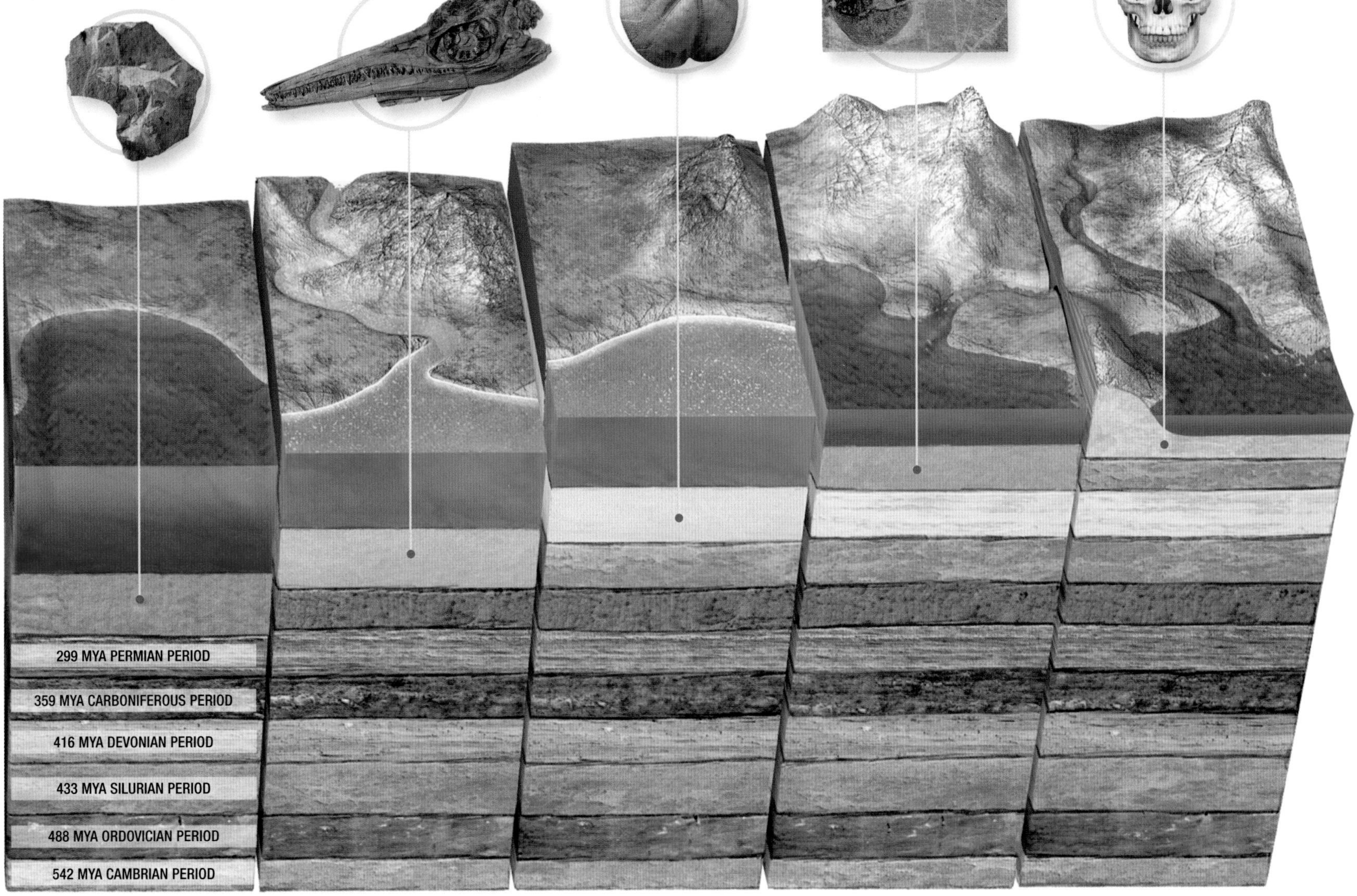

299 MYA PERMIAN PERIOD

359 MYA CARBONIFEROUS PERIOD

416 MYA DEVONIAN PERIOD

433 MYA SILURIAN PERIOD

488 MYA ORDOVICIAN PERIOD

542 MYA CAMBRIAN PERIOD

What are fossils?

Fossils are the remains of once-living organisms, plants, animals, or microbes. They are the key to dating rocks (see pp.14–15), but more importantly they reveal the history of life. Human fossils can show us astonishing detail of ancient anatomy, diet, locomotion, and behavior.

Fossil collections

Over the years, millions of fossils have been collected, studied, and categorized. Despite this vast number of fossils in museums and private collections, most plant and animal species that have existed never met with the very specific conditions needed to make it into the fossil record (see p.18). It would be easy to conclude from this that the fossil record is too poor to say much about the history of life. However, a single well-preserved specimen with a unique morphology (structure) is all that is needed to first identify a new species. We then learn more about a species as further specimens are discovered.

Hard parts

Fossils are nearly always incomplete, representing the hard parts of the ancient organism. Invertebrates such as ammonites and belemnites leave their shells, and vertebrates such as humans and dinosaurs leave their bones. Normally, the muscles, internal organs, and skin rot away quickly and are lost. Bones and shells may be preserved largely in their original form, perhaps with the spaces filled up with minerals.

PILTDOWN MAN FOSSIL FORGERY

Piltdown Man was found in 1912 at Piltdown, England. The partial skull and jawbone were taken to Arthur Smith Woodward at the British Museum of Natural History. He thought that the Piltdown fossil was a human ancestor—with its primitive jaw and modern braincase—and named it *Eoanthropus* ("dawn man"). The Piltdown hoax was only exposed in 1953, thanks to the work of a new generation of scientists. Fluorine dating showed that the skull was ancient, but the jaw was from a modern orangutan.

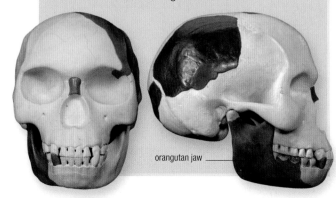

orangutan jaw

NEAR-COMPLETE SKELETON
The partial skeleton fossil of a male Neanderthal about 60,000 years old, from Kebara, Israel, shows nearly all the bones are still in place. Even the tiniest wrist bones have survived.

Body and trace fossils

There are two kinds of fossils. Body fossils show part or all of the body of an ancient animal or plant, whereas trace fossils are remnants of ancient behavior, such as footprints or burrows. Most body fossils are preserved as hard parts that often retain their original internal structure. The normal mode of preservation of a fossil (see pp.18–19) is decay, burial, replacement, and fossilization.

Fossils can be beautiful objects, and they have been collected for millennia because of the mystery about their origin. Early philosophers debated whether such fossils could be real, especially when they found seashells high in the mountains. In medieval times, fossils were the subject of religious debate. Only by 200 years ago did people understand what fossils were—the remains of ancient life and a chance to study its development.

PETRIFIED FOREST
These ancient segments of a tree trunk, from the Petrified Forest National Park in Arizona, USA, preserve all internal detail of the wood. The tree tissues decayed slowly and were replaced by minerals, so every detail of the original structure can still be seen.

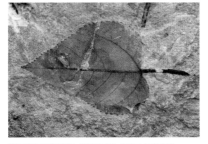

spider preserved in amber

outer surface of coral replicated

carbon imprint of leaf

UNALTERED PRESERVATION
It is unusual for fossils to be preserved unaltered, but insects may be locked in amber (fossilized tree sap) and human bodies may be preserved in peat bogs.

RECRYSTALLIZATION
Many marine animals are composed of calcium carbonate (chalk), which can be altered by acidic waters, compression, or heat, which changes its crystal form.

CARBONIZATION
Plants are made largely from carbon, and so their fossils are often preserved as almost pure carbon. Coal is composed entirely of carbonized plant remains.

REPLACEMENT
This spectacular ammonite fossil has been replaced entirely by iron pyrite. The shell must have been dissolved by acid, leaving a natural mould in the rock, which was later filled with iron sulfide.

IMPRESSIONS, MOLDS, AND CASTS
Many structures, such as this leaf, may leave an impression in mud that can fossilize in the correct conditions. Here, even the color of the dried leaf is imprinted on the rock.

sculptured rib

HUMAN FOOTPRINT
This remarkable trace fossil from Kenya has been dated to 1.5 million years ago. It proves that the early human who made the footprint walked upright, just like modern humans.

The making of a fossil

Fossils come in many forms (see p.17), but the most common preservation sequence consists of death, burial, decay, mineral replacement of tissues, and rock formation. The key time is the first month after death, when soft tissues may be lost and the body may begin to fossilize under the right conditions.

Paleontologists have been interested in taphonomy (the preservation of fossils) for hundreds of years, but the subject has made huge advances through new experimental studies and fine-scale chemical analysis. In fact, many techniques are now shared by forensic scientists, paleontologists, and paleoanthropologists. Modern analytical methods can accurately detect minute quantities of different chemicals—of immense value for the forensic scientist in identifying the perpetrator of a crime, or for the paleoanthropologist in determining the last meal of an ancient body or the time of year when the person died.

DEATH

Today, living coelacanths are found in very deep waters of the Indian Ocean, but in the past coelacanths lived in shallower marine waters. When a fish dies, which may be from either predation or disease, the flesh begins to decompose at once, and often the first stage is for the gut to fill with gas from the chemical reactions created by its last meal. The gas-filled carcass may then float to the surface of the ocean, belly upward. After a day, the carcass may explode as a result of gas pressure or be damaged by scavengers plucking at the flesh. The carcass then plummets to the seabed, and will then often be further scavenged and scattered by fishes, crabs, and gastropods on the seafloor. In some cases, these scavengers consume the bones, but usually they remove the flesh and skin, and scatter the bones.

BURIAL

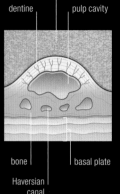

For a complete and perfect fossil, as here, the fish needs to escape being scavenged by falling rapidly into anoxic (oxygen-poor) mud on the seafloor. The lack of oxygen means there are no predators present to scatter the bones, and might also limit the action of microbes. The scales and bony plates of the skull are covered with mud, leaving pulp cavities and tubes (called Haversian canals) in the bone tissue unaffected.

sediment
dentine
pulp cavity
bone
basal plate
Haversian canal

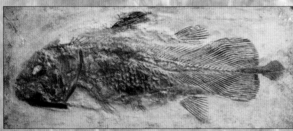

LIVING FOSSIL
This fish is a coelacanth, a primitive kind of lobe-finned fish distantly related to lungfish and land vertebrates. It is sometimes called a "living fossil" because its living relative, *Latimeria chalumnae* (top), which inhabits the depths of the Indian Ocean, is very similar to the ancestral form (bottom), over 250 million years old.

DECAY

Experiments show a definite decay sequence of different body parts. The tissues of a fish range from labile (decay quickly) to recalcitrant (do not decay). Among the most labile tissues are the soft tissues of the head, which would normally decay in a matter of hours or days, even if they are not scavenged for food. In experiments on the decay of specimens of modern lamprey fish, the mouth structures had disappeared within 5 days, the heart, brain, and gill structures in 11 days, the eyes in 64 days, muscle blocks and fins in 90 days, and the gut and liver in 130 days. In the case of fishes, the scales may survive decay, but they are not attached to the skeleton and can be carried away by the lightest water current.

REPLACEMENT

Most uncertain are the processes of replacement that make the skeleton into a fossil. The bones may be compressed by overlying sediment, and any spaces, such as the pulp cavity and the Haversian canals, are filled by sediment or minerals percolating through the tissues. Generally, the fossilized fish bones retain their internal structure, and they are still made of apatite (calcium phosphate), as they were in life. The weight of sediment accumulated above the skeleton can compress the remains, and flatten or distort them.

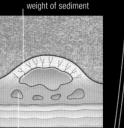

scale's surface and cavities become compressed under weight of sediment

water percolating through sediment picks up minerals in solution and fills scale's cavities

ROCK FORMATION

In some cases, the fossil will not be altered further, and it can reveal fine detail under the microscope. In many cases, however, as the surrounding sediment (mud or sand) turns into rock (mudstone or sandstone) by compaction and cementing of the grains, the fossil may be further compressed, and this may lead to fine-scale mineral replacement occurring within the bone.

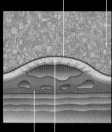

further flattening due to compression

sediment turns into rock

original bony part of scale has been replaced and pores filled with minerals crystallized from fluid

less replacement occurs in dentine at top of scale

FROM FISH TO FOSSIL
The main image shows stages in the fossilization of a coelacanth, but these stages are the same for mammal bones, including those of humans and their ancestors. The visualization reconstructs the sequence through death, burial, decay, replacement, and rock formation—taking us from a complete animal through to a fossil.

IDA THE ANCIENT
The lemurlike *Darwinius*, nicknamed Ida, is an exceptionally preserved fossil from a 50-million-year-old lake bed in Germany. It shows not only the entire skeleton, but also the body tissues and fur, which have been preserved as an oily mass.

Ideal preservation

Modern humans are among the most adaptable animals on Earth and are found in all types of environments. However, only certain kinds of environmental settings are favorable for the preservation of their fossils.

Caves are important fossil locations for two reasons: early humans lived in caves for warmth and protection, and so slept, and sometimes died, there; and caves are often sites of sediment accumulation, so any skeletons will be slowly covered by soil and flowstone (newly precipitated limestone), and remain protected. Another excellent setting is seen in the mixed soils and volcanic ashes of the African Rift Valley, where early humans caught fish in the lakes and foraged for fruit and leaves on the banks. Sediment accumulated over much of the area, and volcanic ash sealed in the bones, providing the ideal conditions for the preservation of some astonishing early human fossils.

LAKE TURKANA
Volcanic ash and lava make up the rock and soil around Lake Turkana, in East Africa. This landscape has been volcanically active for 10 million years, and many very early human remains have been preserved.

LIMESTONE CAVES
Researchers and workers stand inside the entrance to the Liang Bua cave, on the island of Flores, in Indonesia. This is the limestone cave where remains of the "hobbit," the small hominin, *Homo floresiensis*, were found.

Preserved bodies

In some conditions, bodies may undergo natural mummification rather than fossilization. The most famous exceptionally preserved human remains are the so-called "bog bodies" from northern Europe and Özti, the ice mummy. So far, about 700 bog bodies have been found in peat bogs in Denmark, Germany, Holland, Britain, and Ireland. They date from 8000 BC to 1000 AD. Bog bodies are hugely important because they reveal a lot about the diets, clothing, and lifestyles of people who lived thousands of years ago.

Peat bogs are capable of remarkable preservation because they are acidic and anoxic (poor in oxygen). The highly acidic peat acts as a preservative, turning the skin to leather, and often also preserving clothing. When a body is thrown into a bog, the cold water hinders putrefaction and insect activity, which would normally remove the flesh in weeks or months. Sphagnum mosses and the tannin in the bog water add to the preservation because of their antibacterial properties.

leathery skin pulled tight over woman's skull

upper right arm, cut through by the axe blow that killed her

severed right arm, found nearby

bog preservation has pickled the skin and blood vessels

HULDRE FEN WOMAN
This bog body was found in Huldre Fen in northern Denmark in 1879. She had been killed by an axe blow, which also severed her right arm, 2,000 years ago. Huldre Fen woman was wearing a long tartan skirt over a lambskin undergarment, a lambskin shawl, and a tartan cap when she was killed.

THE ICEMAN
Probably the most famous natural mummy is Ötzi, discovered in 1991 close to the border between Switzerland and Italy. Ötzi lived 5,300 years ago, and was clothed in thick fur garments and carrying a copper ax, a flint knife, and a quiver filled with 14 arrows.

BUSY DIG SITE
The Gran Dolina site in Spain is famous for human fossils dating back approximately 800,000 years. Sites like these attract huge attention, and excavation crews may be very large.

Finding our ancestors

The frantic activity of an archeological excavation site may be familiar to many from television programs. Professional archeologists may be joined by volunteers and students, who are directed according to a careful plan, each working slowly down the sediment, cleaning and photographing all finds, before objects are logged and stored.

Digging them up

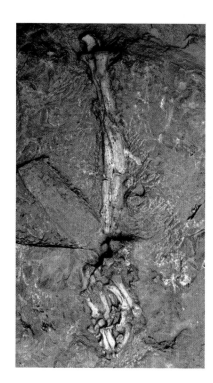

ANCIENT HAND
The photograph shows the fossilized left hand and forearm of an *Australopithecus* species at the Sterkfontein caves, South Africa, as it was first seen by excavators. These remains are the exposed parts of a complete skeleton thought to be around 4.7 million years old.

Every scrap of information must be extracted from an archeological site and the dig team may therefore include experts from many disciplines. Sedimentologists record the soils and rocks; palynologists collect and process soil samples to look for pollen and spores that might identify environmental conditions; palaontologists identify bones found at the site; geophysicists use remote sensing to detect what lies beneath the surface, and to map object finds in 3-D; and experts in radiometric dating can take bone and artifact samples for absolute dating (see p.10).

Archeologists themselves also specialize. Field archeologists are experts at excavating. Various other specialists are called in to help with the finds: cultural archeologists work on the artifacts; osteoarcheologists or physical anthropologists focus on the anatomy of the skeletons; and archeological scientists may carry out chemical analyses of bones and artifacts.

Basic techniques

Modern technology is hugely important to archeologists, but the basics of a successful dig have changed little in 200 years—patience and careful hand work.

Sites are often chosen based on chance discoveries, when perhaps a few bones or tools are washed out on the surface. Archeologists may track these back to a cave or other possible living site, and then an excavation is plotted.

CLEARING AND BRUSHING
The standard tool for carefully digging down through archeological layers is a trowel, although a brush may be used if the sediment is dry and sandy.

Improvements

Three things have changed since the 19th century. First, modern archaeologists can call on mechanical diggers to quickly remove overburden (soil covering the layers of interest), but they must be careful not to damage the buried artifacts. Second, geophysical survey and laser-measuring equipment allow the precise recording of sites and even single objects in three dimensions. Third, there is now an amazing array of paleontological and chemical techniques available in the laboratory for interpreting environments and dating finds.

REPLICAS
Investigators may take casts or make accurate 3-D models of fossils using laser or CT scans. This means that originals are not damaged by handling and data can be easily shared.

FIELD TOOLS
The standard field tools in paleoanthropology have not changed in 100 years—no machine can replace the hammer, scalpel, and brush for precision, close work.

Interpreting the site

Much of the significance of an archeological site becomes evident only in the laboratory. Skulls and skeletons have to be cleaned carefully and reconstructed before they can be interpreted. Chemists may detect chemical residues in utensils that identify foods, or they may carry out isotopic studies on bones to determine diet and where an individual was born. Further, chemical study of the bones may allow estimations of the age of the site.

The interpretation depends also on the site itself and its context. Some sites were occupied on more that one occation in their history, and different occupation levels have to be identified and interpreted. Animal bones, plant remains, tools and artifacts, and soil and rock samples can all be used to build up a picture of the site during different phases of occupation, revealing clues on changing environments, diets, and behavior.

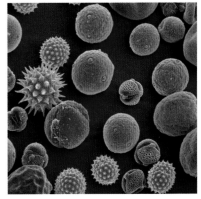

POLLEN
Tiny grains of ancient pollen, viewed here under a microscope, can be identified to the species that produced each grain. This indicates ancient environmental conditions, such as temperature and rainfall.

ROCK STRATA
These layers of sediment tell a detailed story of environmental change. The top layer is volcanic ash, which tops layers of pebbly limestone, mudstones, and sandstones, each bed representing a different shallow marine environment.

THE SITE ORGANIZING AND MAPPING

Mapping the exact position of a find is crucial to dating and interpreting what went on at the site. Archeologists use gridding methods, where 3-foot (1 m) squares are defined by strings, and an artist can quickly locate and record each find in place before it is removed. Once a surface is cleaned, and all artifacts displayed, photographs make a permanent record, and laser survey and LIDAR (light detection and ranging) can be used to measure precisely the complex three-dimensional topography of a surface and each object's location.

1. **Sampling soil core for plant pollen.**
Pollen released by plant species, preserved in sediments, provides clues about past environments and climates.
2. **Microscopic examination of rocks.**
The different proportions of minerals in rocks and ores allow archaeologists to establish the origin of the raw materials.
3. **Linear particle accelerator dating.**
An accelerator allows for the dating of very small samples of material, meaning that less damage is caused to the artifacts.
4. **Analyzing genetic material.**
Extracting and studying DNA from human and hominin remains can help to establish the evolutionary relationships between different hominin species.

Archaeological science

Archaeologists rely on many different techniques to locate and excavate sites, to establish how old they are, and to analyze the artifacts and remains found there. Combining various techniques can help to establish the environments in which people lived, where their belongings came from, their diets, and more about their everyday life.

Archaeologists have a limited window on the past and must make the most of every piece of evidence available. Many different scientific techniques allow us to locate and excavate sites, establish how old they are, and analyze the artifacts and remains found there. New imaging techniques, such as remote sensing from satellites or aerial surveys, allow archaeologists to locate possible sites in remote or difficult-to-access parts of the world from afar. Once located, geophysical surveys are used to investigate any physical structures before decisions are made about how to excavate them.
The age of a site can be determined using a combination of different dating methods, each one appropriate for a particular age range and dating material.

Other kinds of analyses allow archaeologists to understand the environments and climates in which people lived. Particular species prefer certain conditions, so determining which species of mammals, snails, beetles, plants, or pollen occurred at a site can provide clues to past environments and climates. Human bones can be chemically analyzed to determine what kinds of foods people ate, and sometimes even where a person came from. Cultural remains can also be analyzed. The exact composition of stone and metal, for example, can tell us where the raw materials used to make different objects came from, shedding light on how these raw materials and objects were traded and exchanged in the past.

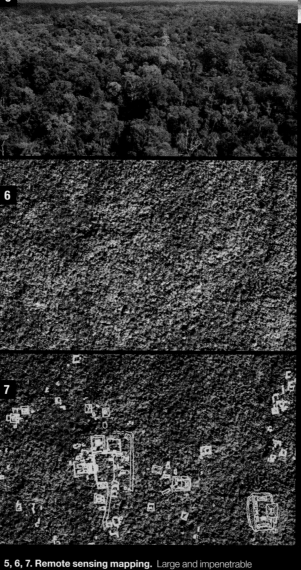

5, 6, 7. Remote sensing mapping. Large and impenetrable landscapes can be surveyed using remote sensing. Remote sensing using reflected or emitted radiation gathered from aerial photographs (5) or satellite images (6) may be used to distinguish between different types of ground cover, vegetation, and structural remains (7).

Genetic evidence

Comparisons of DNA between living people from different parts of the world allow scientists to study the evolutionary relationships between different populations, and even help to show how modern humans spread around the world. This work began with a focus on relatively small portions of DNA, from structures called mitochondria inside cells, as well as the Y chromosome. But now whole genomes—encompassing all the genetic material contained in the cell's chromosomes—are being sequenced and compared.

DNA accumulates mutations over time. When individuals in a species can interbreed freely, these mutations spread evenly through the population. However, if groups are no longer able to interbreed as often—perhaps because they are separated by large distances or mountain ranges—genetic differences between them slowly accumulate. Using estimates of the rate of mutation, it's possible to determine how long ago two species or populations separated (see p.58). For example, the human genome is quite similar to that of the chimpanzee but very different to that of a shark. Similarly, the genes of two people of European ancestry are more similar to each other than to those of a person with sub-Saharan African ancestry, though these differences are tiny—all modern humans have extremely similar DNA.

Analysis of ancient DNA from human remains can indicate how closely individuals buried at a certain site were related, and sometimes even what diseases affected them. DNA can also be extracted from the bones of some extinct species. The sequencing of DNA from Neanderthals has revealed that they interbred with modern humans and contributed genes to modern human populations. DNA has been extracted from even older fossils, dating to hundreds of thousands of years ago, helping scientists to understand more about the hominin "family tree."

As geneticists learn more about how the genome relates to specific aspects of physiology and anatomy, we will be able to tell more about what the genetic differences between modern humans and our ancestors actually mean. Many have already been found to involve the immune system. Some may eventually help to explain human brain development and cognition.

DNA EXTRACTION
A technician drills a Neanderthal bone to extract a DNA sample. The genetic material, extracted and sequenced as part of the Neanderthal genome project, will give insight into human and Neanderthal relationships.

Piecing it together

Excavated skulls and skeletons are usually incomplete, and it may seem as though paleoanthropologists use a great deal of imagination in making reconstructions. Putting together a human skull is the ultimate jigsaw puzzle, but the guesswork is minimal.

Fragmentary finds

There are three steps in piecing together an ancient skull or skeleton. The excavator notes exactly where each fragment lies—close association of pieces gives clues about what fitted to what. This is hugely important when the skull or bone has been broken during recent weathering processes. Then, the pieces are carefully removed and the surrounding soil is sifted to make sure nothing is lost. Finally, the anthropologist uses all available clues in the laboratory, their knowledge of primate anatomy, modern analytical tools, and great patience to produce highly complex reconstructions.

SKETCHING AND NOTING
An excellent visual sense is needed to draw and interpret archeological finds. The ridges and sutures (immovable joints) of every bone are recorded and each element is compared with modern bones to make accurate records and reconstructions.

PUTTING THE PUZZLE TOGETHER
Physical reconstruction is a long and painstaking process, and here a scientist uses stands and clamps to hold the pieces of this 300,000-year-old human skull together.

LUCY RECONSTRUCTION
This reconstruction of the skeleton of "Lucy" (*Australopithecus afarensis*) is based on a partial skeleton found in Ethiopia in 1974. Only parts of the skull and skeleton were preserved, but the symmetry of the skeleton allows right and left elements to be reconstructed.

Filling in the gaps

It is rare to excavate a complete and undisturbed skeleton. Normal processes of decay and scavenging by insects and animals may have damaged and scattered bones before they are buried. Indeed, many bodies will have been damaged at the time of death or shortly afterward. Then, more recent weathering of exposed sites will break up skulls and bones and scatter the pieces widely.

Close study of modern skeletons allows paleoanthropologists to identify often quite fragmentary pieces and attempt to reassociate scattered materials. Modern technology, in the form of computer programs that allow 3-D virtual reconstruction, have revolutionized the reconstruction process: once the isolated and fragile bones have been scanned, they can be manipulated in the computer with great precision and with no risk of damaging the specimens. Gaps can be filled in and distortions that happened during fossilization accounted for and ironed out.

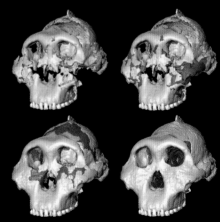

SKULL RECONSTRUCTION
Using medical scanning techniques and computer software, it is possible to virtually reconstruct a skull of *Paranthropus boisei*. Gaps are filled by mirroring from right to left, by filling in narrow cracks, and by comparing the skull to other fossils for the final details.

SKULL FRAGMENTS
These small fragments of bone might seem to be featureless, but through hours of patient work these pieces of fossilized skull from Sima de los Huesos, Spain, were fitted together to form more complete skulls.

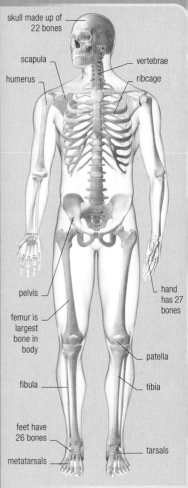

HUMAN SKELETON

skull made up of 22 bones

scapula

humerus

vertebrae

ribcage

pelvis

femur is largest bone in body

hand has 27 bones

patella

fibula

tibia

feet have 26 bones

tarsals

metatarsals

The human skeleton is known in great detail due to medical study, and anthropology gives a detailed inventory of variation within human populations, and between populations worldwide. This is crucial for studies of individual fossil finds, and for comparative studies of ancient populations.

Understanding differences

The human skeleton shows great variation among modern humans, depending on factors such as diet, climate, history, and age. A poor diet may lead to diseases such as rickets, which shows up in bowed legs and short stature. People in cold climates tend to be stockier and shorter than those in temperate climates. We also inherit aspects of our stature and body shape from our parents and grandparents.

Males and females also show skeletal differences. Females have a shorter and wider pelvis than males because of the requirements of giving birth. Men tend to have thicker and longer limb and finger bones, reflecting their more muscular arms and legs. In some archaeological sites, men also seem to show more wounds and injuries, perhaps reflecting a more violent

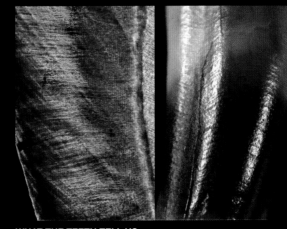

WHAT THE TEETH TELL US
This composite image clearly shows growth lines inside (left, diagonally running lines) and on the outside (right, horizontal curved lines) of a Neanderthal tooth. Counting and measuring these lines helped determine that the child was

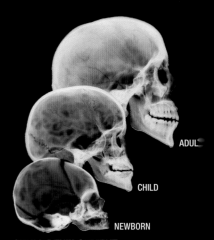

ADUL

CHILD

NEWBORN

SKULL DEVELOPMENT
X-rays of a newborn, child, and adult skull show changes in shape and structure. Gaps between a newborn's skull bones become sutures by adulthood, and the child's teeth

Bringing bones to life

For over 150 years, anthropologists have tried to visualize what human ancestors looked like. Early efforts were not only cartoon-like, but also reflected considerable preconceptions—for example, about how "primitive" they thought a particular hominin might have been.

Reconstructions

Paleoanthropologists now have a remarkable battery of scientific methods to aid in reconstructions, including detailed knowledge of human anatomy. The same techniques are used by forensic scientists to reconstruct the faces of missing people from their skulls, and the reconstructions often show an uncanny resemblance to the crime victim once they have been identified. Reconstructing a head from fragmentary material is a time-consuming and costly process, but scientific interest in human evolution is so intense that the expenditure of time, effort, and money is justified.

3-D scanning

After the reconstruction of fossil remains (see pp.26–27), a three-dimensional scan is produced. 3-D scanning is becoming an increasingly important part of archaeology and anthropology. It is now possible to take millions of measurements of objects and sites, recording them at such high resolution that they can be digitally recreated with a very high level of accuracy. This allows scientists to study them wherever they are in the world. Since access to originals is often no longer necessary, scanning also helps to preserve fragile finds, such as fossils or even whole sites that may be threatened, for example, by excessive tourism, climate change, or even war.

plastic copy of human ear

muscle structure molded with clay

INTO THE SCANNER
A modern human skull and fragments of a Neanderthal skull are passed through a computerized tomography (CT) scanner. This images many "slices" through each object, forming the basis for a 3-D reconstruction.

STRESS AND STRAIN
This finite element model of a cranium of *Australopithecus africanus* is based on the specimen in the background. Finite element analysis is an engineering technique, used here to color-map the magnitude of strains associated with biting on a hard object. This gives direct evidence about this species' diet.

pegs used to indicate skin thickness

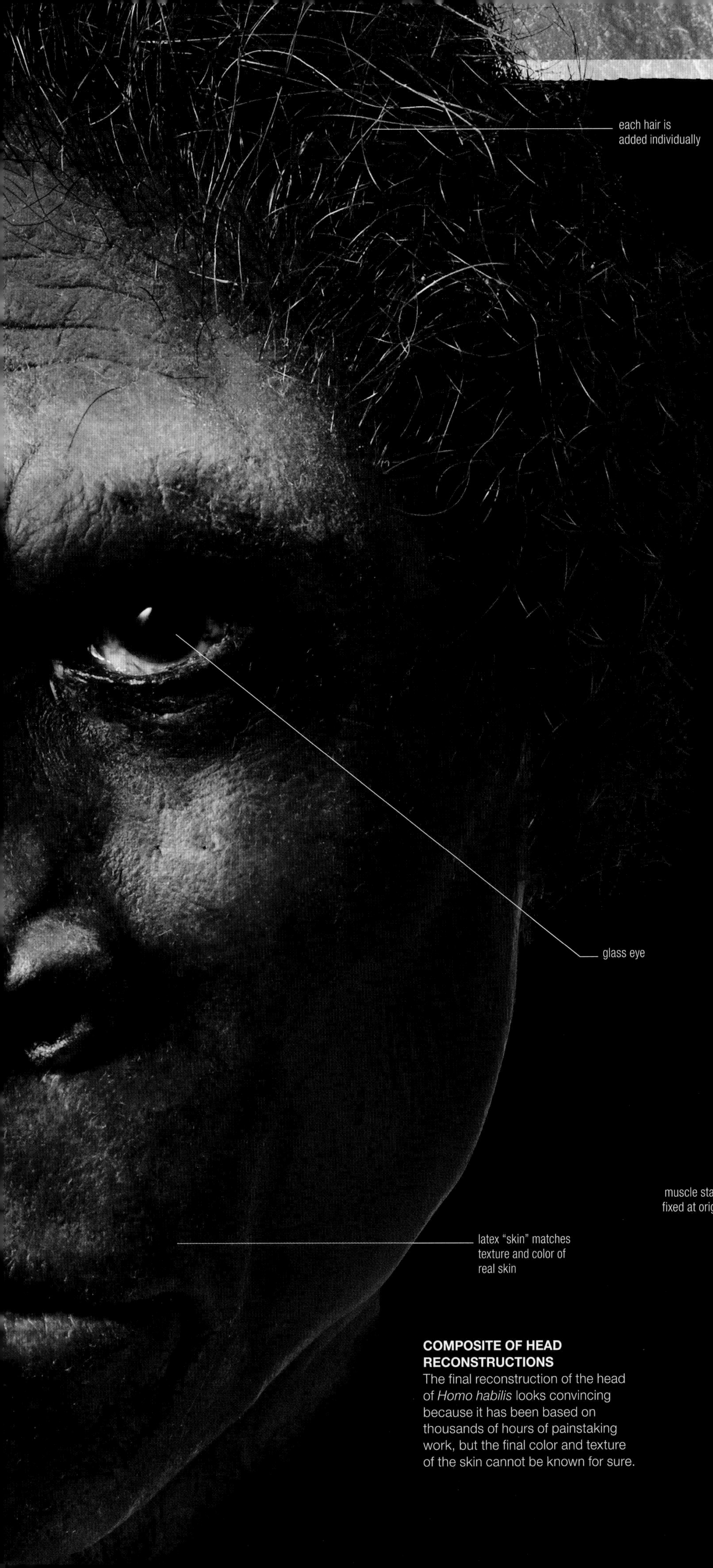

each hair is
added individually

Putting flesh on bone

The hardest part of the reconstruction is to add flesh
to the skull. The anatomist uses two techniques that are
common in forensic science as well, namely a thorough
knowledge of the soft tissues of the human head based
on dissection and the use of clues in the skull to
determine the shape and orientation of external features.

Modern humans and chimpanzees (see pp.54–55)
share all the main facial muscles, and so it is reasonable
to assume that ancient hominins, such as *Homo habilis*,
also had those muscles. These muscles are added using
modeling clay. Rough patches on the bone can give clues
about how massive certain muscles might have been.

Surface details, such as skin and eye color and hair
texture, must be modeled using comparisons with
modern humans and other living primates. However,
objective reconstructions reveal a great deal about what
our ancestors would have looked like and provide
important insights into the evolutionary relationships
between them. They also enable us to literally come
face-to-face with our ancestors, bringing the past to life.

glass eye

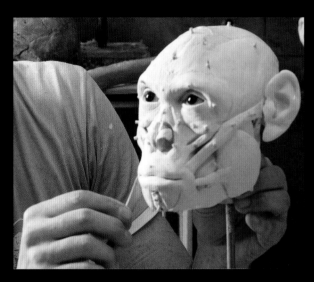

APPLYING LAYERS
Here the main muscles have been modeled in clay. Ears
and noses are difficult to reconstruct because they have no
bony parts. Comparison with modern apes and humans,
and the shape of the nasal opening in the skull can show
the orientation and likely size of the nose and ears.

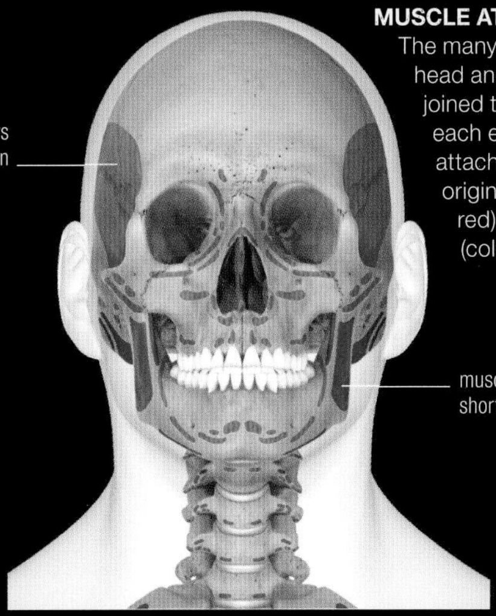

MUSCLE ATTACHMENTS
The many muscles of the
head and neck are
joined to the bones at
each end through
attachments known as
origins (here colored
red) and insertions
(colored blue).

muscle stays
fixed at origin

muscle contracts and
shortens at insertion

latex "skin" matches
texture and color of
real skin

**COMPOSITE OF HEAD
RECONSTRUCTIONS**
The final reconstruction of the head
of *Homo habilis* looks convincing
because it has been based on
thousands of hours of painstaking
work, but the final color and texture
of the skin cannot be known for sure.

Reconstruction

Making accurate anthropological models is specialized work, and there are now several studios around the world where highly skilled technicians produce lifelike and often quite startling 3-D reconstructions of ancient hominin skulls.

The Dutch paleoartists Adrie and Alfons Kennis specialize in carrying out all the stages in reconstruction of hominin skulls. This is a skilled process that requires knowledge, artistry, and imagination. The quality of reconstruction work depends on years of background study, but some features, such as eye color, can never be known for certain.

The example shown here shows the steps in reconstructing a head, and the completion of a Neanderthal (see pp.148–57) head. The focus here is on the molding and casting process and the artistic finish. Molding is a common process in reconstuctions—it is used to make replicas of an important skull or bone for study purposes. Modern synthetic rubbers can record every striation and crack in an original model or bone. In the reconstructions studio, the cast is essential to allow the artists to produce a head in fleshlike, soft rubber.

1. Studying the skull. The Kennis brothers study a model of the skull, noting roughnesses on the bone, and the exact angle and shape of the nasal opening.

2. Pegs added for skin measurements. Humans have predictable flesh and skin thicknesses—thin over the forehead for example, and thicker over the cheeks.

3. Muscle structure sculpted. The muscles are added using modeling clay, and based on the anatomy of humans and apes.

4. Layers of plastic clay added. Skin is modeled with thin sheets of clay.

5. Face shaped and texture added. The head is finished by marking skin textures and shaping the lips.

6. Silicone rubber painted onto head. In order to make casts, the sculpted head is molded first by painting silicone rubber over the model. The rubber sets to form a flexible mold.

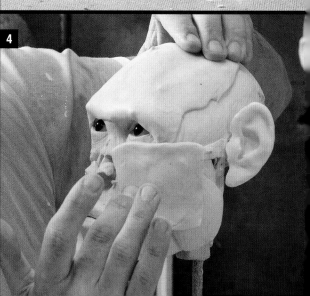

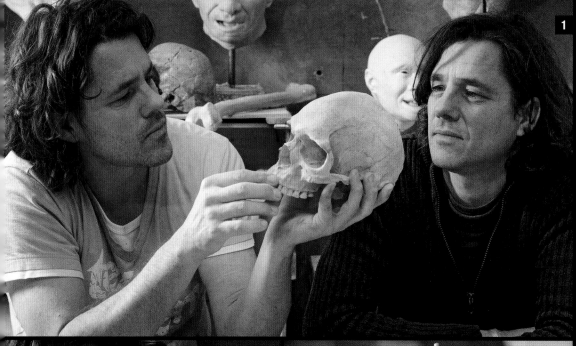

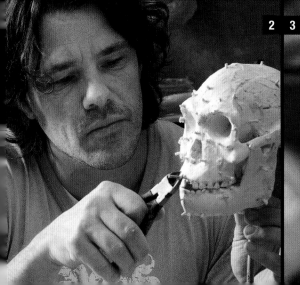

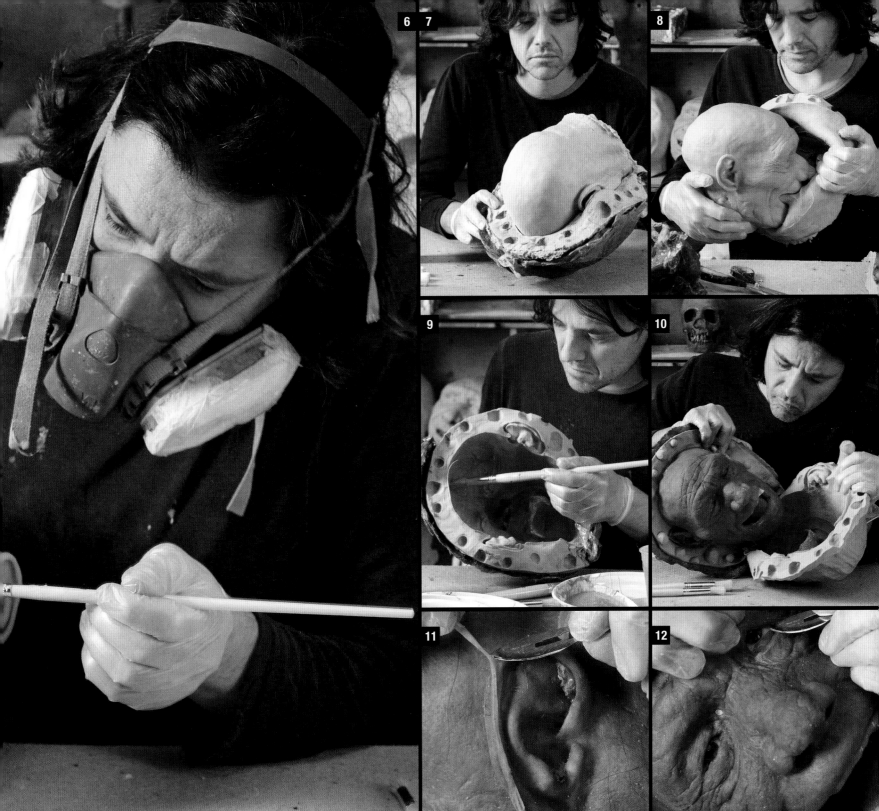

6 **7** **8**

9 **10**

11 **12**

5

13

7. **Mold is built up.** A support mold is added for strength.

8. **Dry mold taken off head.** Once the silicone has set, the mold is removed from the model.

9. **Mold is painted.** The layers of pigment that give the skin its color are brushed on to the inside of the mold.

10. **Head broken out of mold.** Casting plastic is poured into the mold, it sets, and the mold is then peeled off.

11. **Ears trimmed and finished.** Molding edges have to be trimmed off.

12. **Eyes opened and finished.** Glass eyeballs are added.

13. **Hairs added to scalp.** Each hair has to be inserted separately to create a realistic appearance.

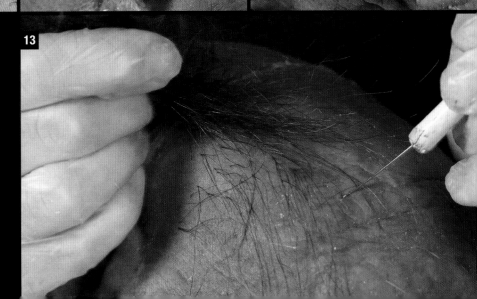

Interpreting behavior

Studying the anatomy of our ancestors can not only give us clues to what they looked like, but can also provide insight into how they behaved. Artifacts such as tools, paintings, and other forms of art may also be revealing about their behavior.

embedded microlith (tiny blade)

Tools

Archaeologists have long identified increasing complexity in the styles of tools used by early humans. Paleolithic ("ancient stone"), Mesolithic ("middle stone"), and Neolithic ("new stone") tools show increasing sophistication, from rough pebbles, through partly shaped flints, to subtly chipped and polished flints. At first, the sequence could be arranged in relative order, but not dated. Some of the Paleolithic tools and weapons were so crude that many doubted they had even been shaped. Now, with accurate dating techniques, careful comparison between sites, and the efforts of modern-day flint-knappers, the detailed characteristics of different tool types, or modes, can be defined and sequenced.

MODE 1
The simplest kind of pebble tool from Olduvai Gorge, Tanzania, dated from 2.6–1.7 million years ago.

MODE 2
A large Paleolithic handaxe, classed as part of the Acheulian culture, which lasted from 1.7million to 100,000 years ago.

MODE 3
Tools from 300,000–30,000 years ago show evidence of complex, preplanned sequences of knapping to produce a sharp, preshaped flake.

MODE 4
A tool from 45,000–35,000 years ago shows assured knapping that produced a sharp blade.

MODE 5
This mode is distinguished by finely worked microliths, which were hafted onto spears and handles.

MODE 6
During the Neolithic, stones were ground and polished to form standardized tools.

sharp edge

AX WITH HANDLE
A complex tool, such as this ax, shows that its maker could plan and execute complicated actions.

The brain

Brain tissue does not fossilize, and so cannot be studied directly. Fortunately, mammalian brains fit fairly tightly inside the cranium and this allows the brain size of fossil mammals to be estimated by measuring the cranial capacity. A traditional way of doing this was by pouring in mustard seeds, then measuring the volume of the seeds, but now 3-D scans can produce more accurate measurements. However, large animals also have large brains, so intelligence may be better measured by relative, not absolute, brain size. As the sequence below shows, the size of brains relative to the mass of the whole body has changed substantially over time. The complexity of tools produced also indicates the capability of planning and problem solving, another measure of intelligence.

BRAIN SIZE AS BODY MASS
The proportion of the body mass composed by the brain (below) rises from a modest 1.2 percent in *Australopithecus afarensis* to 2.75 percent in modern *Homo sapiens*.

1.2%	1.36%	1.17%	1.56%	1.58%	1.46%	1.69%	1.98%	2.75%
3.7–3 MYA	**3.3–2.1** MYA	**2.3–1.4** MYA	**2–1.2** MYA	**2.4–1.6** MYA	**1.8–0.03** MYA	**600–450** KYA	**430–40** KYA	**300** KYA
AUSTRALOPITHECUS AFARENSIS	*AUSTRALOPITHECUS AFRICANUS*	*PARANTHROPUS BOISEI*	*PARANTHROPUS ROBUSTUS*	*HOMO HABILIS*	*HOMO ERECTUS*	*HOMO HEIDELBERGENSIS*	*HOMO NEANDERTHALENSIS*	*HOMO SAPIENS*

Learning to speak

Anthropologists continue to debate when speech first developed among our ancestors. The larynx is composed of soft tissue and does not fossilize, so our ancestors' speech anatomy must be reconstructed from other lines of evidence. The hyoid bone, from which the larynx hangs, does sometimes survive and provides some clues, while the size and shape of the mouth and the size of the holes in the skull through which the nerves controlling speech pass also reveal detail regarding possible speech. But it is impossible to know the detailed vocal-tract anatomy of our ancestors. Because even modern human language is not always spoken, and human vocal anatomy is quite similar to that of apes, it seems likely that the key to mastering speech is not our anatomy, but lies in the brain instead.

HYOID BONE
The hyoid is a U-shaped bone that anchors the tongue and from which the larynx is suspended. It is not attached to any other bones.

greater horns of hyoid may be felt, just underneath jawbone

Culture

The brain size of early humans can be assessed from their skeletal remains, but to understand behavior, one must discover how these abilities might have been expressed. "Culture" is the general term for patterns of human knowledge, belief, and behavior that arise from the capacity for symbolic thought and social learning. Some aspects of culture are found in the manufacture of useful tools and weapons, but human culture also relates to art and the development of abstract and symbolic thought—creating objects that appear to have no useful function, but have profound symbolic meaning. The most obvious manifestations of early human culture are the remarkable cave paintings of Europe, some dating back up to 32,000 years. The oldest expression of art is still debated—perhaps some scratched pieces of red ocher, or even the Makapansgat pebble, which was not manufactured, but perhaps kept because it represented a face.

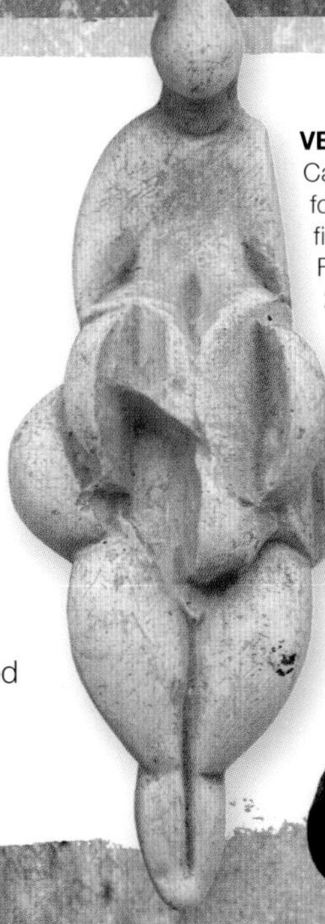

VENUS FIGURE
Carved stylized female forms, such as this figurine from Lespugue, France, (24,000 to 26,000 years old) are common in prehistoric sites.

POSSIBLE FACE
Australopithecus is thought to have recognized this pebble as a representation of a face.

natural markings

carved pattern on red ocher

CARVINGS
This engraving was found at Blombos Cave, South Africa. At around 77,000 years old, it is one of the oldest known complex images.

CAVE PAINTINGS
This famous and evocative example of prehistoric art, showing aurochs (wild cattle) and a small, upland horse, was painted on the walls of the Lascaux cave, in southwestern France, some 17,300 years ago.

PRIMATES

Primates first appeared just under 65 million years ago, after the age of the dinosaurs. The early primates lived in tropical rain forests. They were arboreal and had grasping hands and feet. As they evolved, primates became better-adapted to life up in the trees, developing binocular vision to enable them to judge distance with great accuracy. Today, there are more than 400 species of living primates, the majority of which are still most at home in the trees, but some species, including humans, now live on the ground.

Evolution

Evolution forms the basis of our understanding of modern biology, and the study of humans. The current model of evolution was presented in 1859 by Charles Darwin, who showed that species diverged from each other, thus giving a "tree of life" spanning all of geological time.

What is evolution?

Evolution is the process by which organisms change over the course of generations. Evolution is also creative, in that one ancestor can give rise to many descendants. For example, one of the first known birds, *Archaeopteryx*, which lived 150 million years ago, is close to being the ancestor of all 10,000 modern species of birds. So, the combination of a unique set of adaptations—feathers, wings, lightweight skeleton, excellent eyesight—were the source of a remarkable evolutionary radiation as the first birds exploited all the new opportunities afforded to them by their ability to fly. The

key to evolution is natural selection, whereby members of a species that are best adapted to their local environments survive, breed, and pass their special characteristics to their young; the fittest young thrive and breed, and eventually an entirely new species is formed.

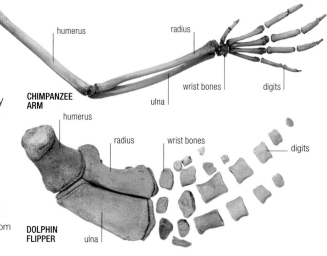

CHIMPANZEE ARM
humerus · radius · ulna · wrist bones · digits

DOLPHIN FLIPPER
humerus · radius · ulna · wrist bones · digits

ARMS AND FLIPPERS
Even though a dolphin flipper looks very different from a chimpanzee arm, and they are used for different purposes, their basic anatomy is the same, showing they arose from a common ancestor millions of years ago.

Genes and inheritance

Darwin knew that evolution could work only if there was inheritance. He could not have had a knowledge of modern genetics, but during the 20th century, it became clear that the genetic code he sought was contained in the chromosomes inside the nucleus of almost every cell of every living organism. Each human cell contains 20,000–25,000 genes, each of which has coded instructions for specific characteristics. The code is mainly in the form of a molecule of DNA, which includes four chemicals, called bases, arranged in pairs. Every gene is encoded by a specific sequence of base pairs.

chromosome formed by one molecule of DNA

DNA strand

strands linked by base pair

DNA molecule forms a spiral (a double helix)

DNA
Deoxyribonucleic acid (DNA) is the basis of the genetic code, coiling to form chromosomes. The coil can unravel into two separate strands, each of which acts as a pattern for a new piece of DNA, and this process carries genetic information from one generation to the next.

Adaptability

The key to evolution is that organisms are variable. Nothing is fixed. This is clear if you look around a room of people—some have black hair, some red, some are tall, others short. Normal variation in physical characteristics in a species can be very large. Adaptations are the features possessed by organisms that suit them for a particular function. So, primates have evolved binocular vision (see right) and large brains for dealing with their varied forest environments. Many primates have long, powerful arms and grasping hands and feet for holding on to branches and for swinging through the trees. The prehensile (grasping) tail found in some monkeys is an adaptation for even better movement through the trees. Adaptations are constantly changing as the environment in which a species lives changes. If the temperature becomes colder, individuals with longer fur may have an advantage and become more common.

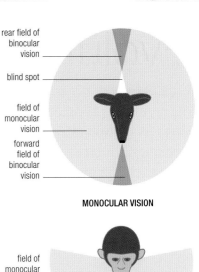

rear field of binocular vision

blind spot

field of monocular vision

forward field of binocular vision

MONOCULAR VISION

field of monocular vision

field of binocular vision

BINOCULAR VISION

FIELD OF VIEW
Primates have forward-facing eyes, giving a wide overlap of the visual fields. This binocular vision enables them to judge distance with great accuracy, such as when jumping from tree to tree. Prey animals, such as deer, have eyes on the sides of their head, giving a very wide, but mostly monocular, field of view.

GEOGRAPHIC VARIATION
The Siberian tiger (left) has thicker fur than the four southern subspecies of the tiger, such as the Sumatran tiger (below), which is the smallest and darkest of them all and may even be a separate species.

What is a species?

A common, working definition of a species is that it's a discrete population of organisms that doesn't naturally interbreed with other groups. By this definition, there are probably more than 10 million species on Earth today, including about 5,000 mammals, of which about 435 are primates. However, every individual within a species is different, and genomes evolve over time. How different can an individual or group be before it should be considered a separate species? Members of different species can still interbreed if they are not yet too genetically distinct. Some species interbreed only under human control; for example, a mule or hinny is the offspring of a horse and a donkey but is sterile. However, other species interbreed successfully in the wild, and we now know that our species *Homo sapiens* did so with Neanderthals and other archaic human species.

Classification

Classification, or taxonomy, is the science of identifying organisms and arranging them in groups according to their evolutionary relationships. Modern classification methods seek to reveal the common ancestor or ancestors of all life on Earth.

SHARKS RAY-FINNED FISHES AMPHIBIANS TURTLES SNAKES AND LIZARDS CROCODILES BIRDS MAMMAL

COMMON ANCESTOR
The animal groups in this cladogram are all related to the first vertebrate—their common ancestor—which appeared about 540 million years ago. The branching patterns are a result of divergent evolution, forming a family tree.

EARLY VERTEBRATES

Types of classification

Early systems of classification grouped organisms according to their general resemblance, and the Swedish botanist Carl Linnaeus (1707–1778) created the system that is still in use today. He set up formal categories on the basis of shared morphological features (form and structure), creating a hierarchy of increasing inclusiveness that ranges from species to kingdom. Since the early twentieth century, it became more widely accepted that classification should be based on the evolutionary relationships between organisms. This "phylogenetic" approach places organisms in groups called clades, based on their morphology and genetic characters. It assumes that a characteristic shared by just one group of organisms means they have a closer evolutionary relationship to each other and a more recent ancestor in common. Phylogenetics (or cladistics) has led to many changes in the classification of a wide range of organisms. For example, birds are now classified as a group of dinosaurs.

Linnaeus chose Latin as the universal language for his classification system and it is still used by most taxonomists today. Every species has a unique two-word Latin name, describing the genus and species. For example, all humans, including fossil humans, carry the generic name *Homo*, but only modern humans are referred to as *Homo sapiens*, or "knowing man."

Primate family tree

The longer that two groups of organisms have been separate, the greater the difference in their DNA, and it has been found that, with due caution, the time since the two groups split from a common ancestor can be calculated. This concept is known as the molecular clock (see p.29 and p.58). However, different tests produce slightly different results and estimates of the date of the split may differ by as much as several million years. The primate fossil record can be used to give a minimum time for the divergence of two groups, by using radiometric dating to calculate the age of fossils that are morphologically distinct, thus helping to fine-tune the clock.

Humans' closest relatives are the chimpanzees, from whom they diverged between 10 and 7 million years ago, and are only distantly related to galagos and lemurs. The common ancestor of all primates lived in the Cretaceous.

FAMILY CONNECTIONS
This family tree of the major groups of primates is based on the latest DNA evidence. Primates began diversifying at the beginning of the Paleocene, but the present-day families of primates arose millions of years later.

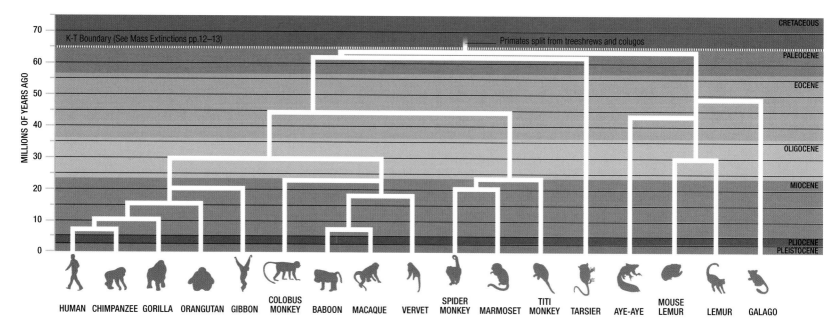

MILLIONS OF YEARS AGO

K-T Boundary (See Mass Extinctions pp.12–13)

Primates split from treeshrews and colugos

CRETACEOUS
PALEOCENE
EOCENE
OLIGOCENE
MIOCENE
PLIOCENE
PLEISTOCENE

HUMAN CHIMPANZEE GORILLA ORANGUTAN GIBBON COLOBUS MONKEY BABOON MACAQUE VERVET SPIDER MONKEY MARMOSET TITI MONKEY TARSIER AYE-AYE MOUSE LEMUR LEMUR GALAGO

What is a primate?

Primates are difficult to classify as there is no unique characteristic that defines them. Some characteristics are shared by other animal groups, while other special features are not found in all primates. In addition, features that relate to soft tissue structures rather than bones, or are behavioral, are of no use when trying to identify a fossil primate. So, a combination of characteristics must be used. Their comparatively unspecialized morphology and highly adaptive behavior has enabled primates to live in a fairly wide range of habitats (and humans now occupy every continent). Although not all species are arboreal, all living primates retain some adaptations for climbing in trees.

BEING FEMALE

As placental mammals, primates give birth to live young and feed them with milk produced by mammary glands. Most primates, like this rhesus macaque, have one pair of mammaries on their chest, but lorises, lemurs, galagos, and tarsiers may have two or three pairs, although in some species not all pairs are functional.

one pair of mammary glands

COMPLEX VISUAL SYSTEM

Forward-facing eyes provide binocular vision, enabling depth perception. Most primates have good color vision; they can distinguish blue and green-yellow hues, and quite a few can distinguish red, too.

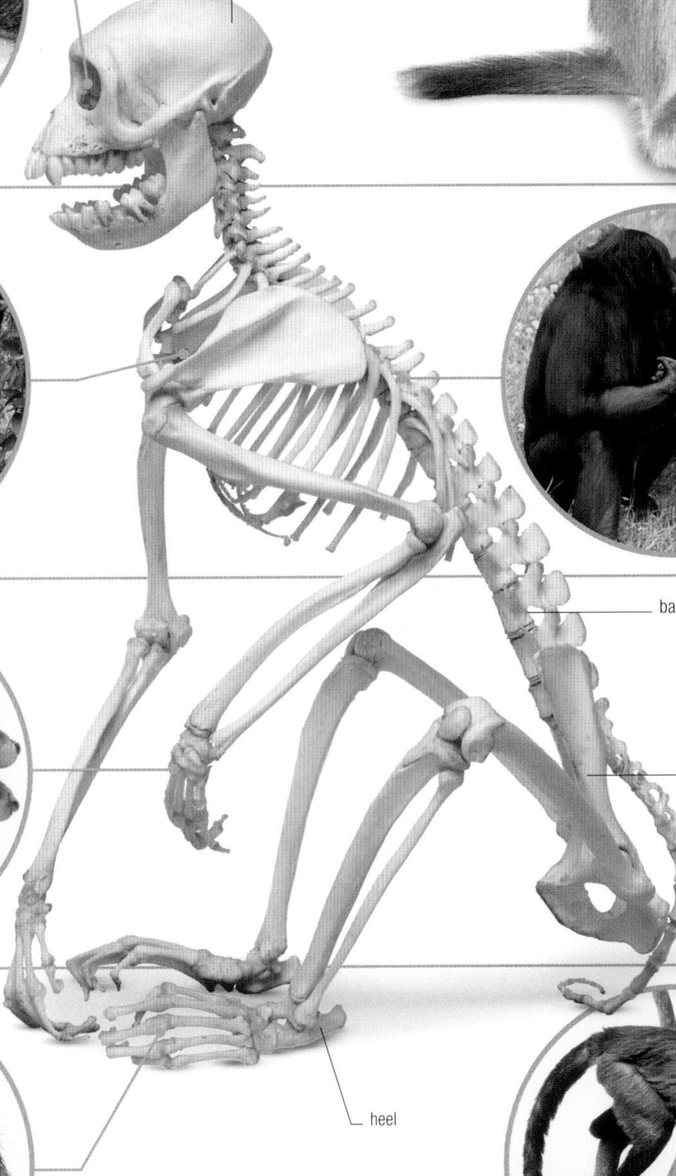

cranium containing large brain in comparison to body size

HIGH DEGREE OF SHOULDER MOBILITY

The apes, and particularly gibbons, such as the siamang shown here, have extremely mobile shoulders, allowing a high degree of movement in all directions.

TORSO HELD UPRIGHT

Several groups of primates will rest and move with upright torsos, and this is a special characteristic of apes. All apes habitually sit upright, as shown by these bonobos, and they often stand upright.

backbone

GRASPING HANDS AND FEET

The hands and feet have five digits. They are modified for grasping and have very sensitive tactile pads, enabling objects to be held with strength and picked up with precision.

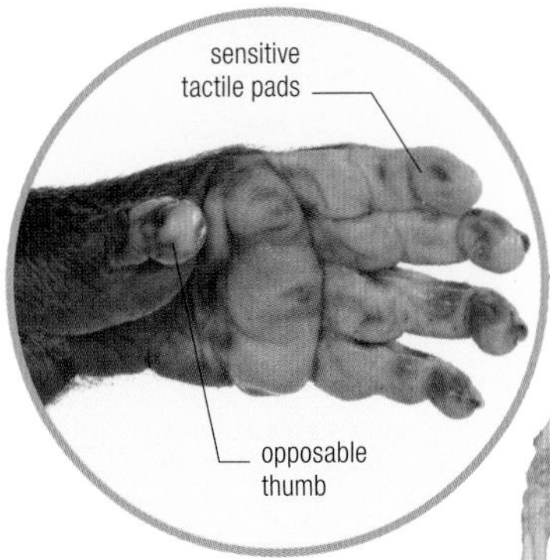

sensitive tactile pads

opposable thumb

PENDULOUS PENIS

All male primates have a pendulous penis and testes that descend into a scrotum. However, there is a marked variation in the length of the penis among primates, and in some species it is brightly colored.

pelvis

FLAT NAIL ON INNERMOST TOE

Some primates have claws on some digits, but most primates have flattened nails to protect the sensitive tips of the fingers and toes. All living primates, except some orangutans, have a flat nail on the big toe.

flat nail on big toe

heel

MONKEY TAIL

Some primates, like this capuchin monkey, have a prehensile (grasping) tail, which they use as a fifth limb when climbing in trees. Others have long tails; some monkeys have very short tails, and apes have no tail at all.

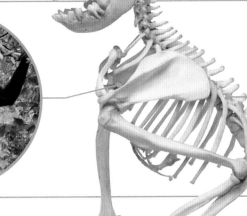

The first primates

Except for the extremely primitive *Purgatorius*, which lived about 65 million years ago in North America, most of the Paleocene primates, the plesiadapiforms, were much too specialized to have been ancestral to modern primates. The primates of modern aspect became abundant only later, during the Eocene.

The Earth in the Paleogene

At the beginning of the Paleogene, in the Paleocene and Eocene epochs, much of the Earth's climate was warm and equable, not strongly seasonal, and it seems likely these types of conditions were widespread across what is now the temperate zone. During the course of the Eocene, the climate became progressively warmer, and the higher temperatures were accompanied by higher rainfall. Even at high latitudes, much of the vegetation about 50 million years ago was clearly tropical rain forest, with plants having such characteristics as "drip tips" (the long

downward-pointing tips of the leaves, allowing rainwater to drip off them), and lianas (climbing, woody vines) were widespread. This implies that the average annual temperature was at least 77° F (25° C), and the patterns of the growth rings in the trees of the period reveal only slight seasonal variation. Fossil remains of mangroves and Nipa palms indicate some areas were swamp forest. The flora was most like that of present-day Southeast Asia, and it was in such an environment, typical of most primates today, that the earliest primates evolved.

SUBTROPICAL SEA
There were no polar ice sheets and the seas around the world were warm in the early Paleogene, even in the high Arctic and Antarctica.

PALEOGENE WORLD MAP
Dry land joined Europe to North America, and the distance between South America and Africa was half of what it is today. The Tethys Sea still separated Africa from Europe and Asia.

KEY
- ANCIENT LANDMASS
- MODERN LANDMASS

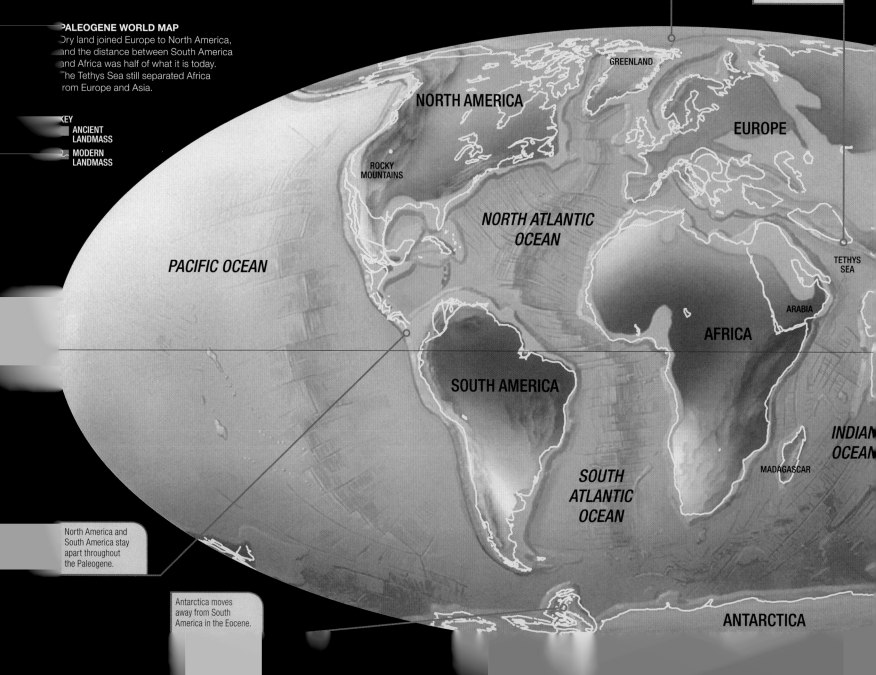

Africa and Asia are brought together as the Tethys Sea closes.

GREENLAND

NORTH AMERICA

EUROPE

ROCKY MOUNTAINS

NORTH ATLANTIC OCEAN

PACIFIC OCEAN

TETHYS SEA

ARABIA

AFRICA

SOUTH AMERICA

INDIAN OCEAN

SOUTH ATLANTIC OCEAN

MADAGASCAR

North America and South America stay apart throughout the Paleogene.

Antarctica moves away from South America in the Eocene.

ANTARCTICA

Early primates

The plesiadapiforms (such as *Dryomomys*, see far right) are almost the only primates known until the end of the Paleocene, 56 million years ago. Their molar teeth were typically primate, but they had small brains, and long, deep faces. Most had claws on all their digits, but *Carpolestes* had a divergent big toe (meaning it splays out from the foot) with a flat nail. Most plesiadapiforms had large, ever-growing incisors, and had lost their canines or premolars, so were already too specialized to be ancestral to modern primates. Some genera were widespread in Europe and North America, others were more restricted in distribution. The first representative of the group from which the ancestors of modern primates arose is *Altiatlasius*, from the Late Paleocene of Morocco, although it is known only from its teeth. Shortly afterward, at the beginning of the Eocene, fossil strepsirrhines (adapids; see right) and haplorrhines (omomyids; see below right) appeared and rapidly began to diversify.

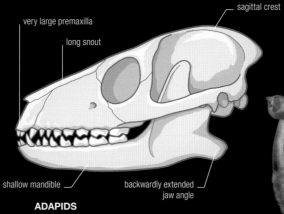

very large premaxilla

long snout

sagittal crest

shallow mandible

backwardly extended jaw angle

ADAPIDS
Adapids and their relatives include the ancestors of modern strepsirrhines (lorises, galagos, and lemurs). Their fossils have been found in Europe, North America, Africa, and Asia, and they flourished throughout the Eocene (one group continued into the Late Miocene). They had fairly small eye orbits and were probably diurnal.

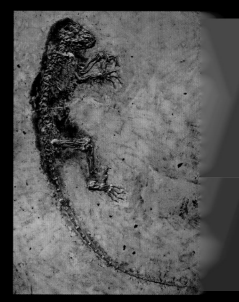

very large orbits

small wedge-shaped nasal bones

large incisors

downwardly extended jaw angle

short, deep mandible

OMOMYIDS
The omomyids and their relatives lived in North America, Europe, Africa, and Asia during the Eocene and Oligocene. They are the earliest haplorrhines, and are perhaps ancestral to tarsiers, monkeys, apes, and humans. Some omomyids had very large orbits (as above) which suggests that they were nocturnal. Their teeth indicate that some omomyids ate fruit while others had a more insectivorous diet.

DRYOMOMYS
First described in 2007, *Dryomomys* belongs to the micromomyids, a family of plesiadapiform primates from the Late Paleocene of North America. It had typically primate grasping hands and feet, but was still primitive with a small brain. It was about the size of a mouse and lived up in the branches of trees, feeding on flowers and fruit.

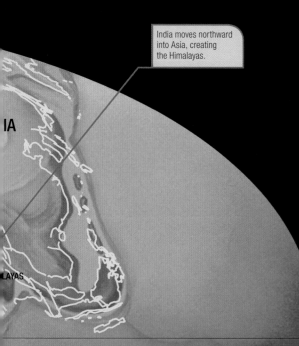

India moves northward into Asia, creating the Himalayas.

IA

LAYAS

AUSTRALIA

TROPICAL RAIN FOREST
During the Paleocene and Eocene, tropical rain forest was the predominant vegetation over much of the world's land masses, even in what is now the temperate zone.

DARWINIUS SKELETON
The discovery of *Darwinius*, represented by a juvenile skeleton from the Eocene, was announced with great fanfare as a "missing link" in the human line in 2009, but many scientists dispute this.

Lorises and galagos

The lorises of tropical Asia and the related pottos of tropical Africa move slowly through the trees on all fours with an inconspicuous gliding motion, tightly grasping branches and vines with specialized hands and feet. The galagos of sub-Saharan Africa, by contrast, cling upright to stems and tree trunks, jumping from one to another with the aid of their long feet and hind legs. Despite their differences in locomotor adaptations, they are very closely related and together with lemurs form the strepsirrhine suborder, although the two groups separated about 50 million years ago. Like lemurs, they have a toothcomb and toilet claws (see below). They live in the same types of habitats as monkeys, but avoid competition by coming out at night, whereas most monkeys forage by day, and they have a better sense of smell than other primates (as do lemurs).

SENEGAL BUSHBABY
Like all galagos, the Senegal bushbaby has large ears that can move, and acute hearing. Its large eyes have a reflective layer that helps it see at night.

SLOW LORIS
Lorises become active at sunset, climbing through the rain forest canopy in search of shoots, fruit, gum, and small animals. The very short tail of the slow loris is usually hidden by its dense fur.

Lemurs

Lemurs are confined to the island of Madagascar, where they have evolved in isolation from primates on the African mainland for at least 40 million years. They have diversified to occupy many niches, with about 100 living species, ranging in size from 1 oz (30 g) to 15 lb (7 kg). A dozen other species became extinct as a result of human activity in the last few centuries, including some giants up to the size of a great ape. Some lemurs are quadrupedal, others are "vertical clingers," leaping upright from tree to tree. Some are nocturnal, but, because there are no monkeys on the island with which to compete, some lemurs are active by day. Nearly all have a toothcomb, in which the lower incisor and canine teeth are pressed together, like the teeth of a comb, and are forward-angled. It is used either for grooming fur or for scraping gum from bark. The toilet claw, a specialized nail on the foot, is also used for grooming.

RING-TAILED LEMUR
This lemur moves by day, on the ground and in the trees of dry forests. It lives in troops of up to 30 individuals, in which females dominate males. Daughters tend to remain with their mothers, but they have to fight for their position within the troop.

RUFOUS MOUSE LEMUR
Mouse lemurs are the smallest living primates, with this species weighing just 1⅖–1⅗ oz (40–45 g). It feeds on fruit, gum, and insects in the rain forests of eastern Madagascar.

AYE-AYE
The solitary, nocturnal aye-aye is the most distinctive of Madagascar's lemurs. It uses its elongated middle finger to locate and extract insect larvae from dead trees.

Tarsiers

Tarsiers look superficially rather like galagos, but they differ in many ways. Unlike galagos, they do not have the reflective layer behind the eye that aids vision at night (so, no "eyeshine"); the nose is not naked and moist; they have a continuous upper lip, and they do not have a toothcomb. They share these features with monkeys, apes, and humans, with which they belong in the haplorrhine suborder, but the two groups split over 60 million years ago (see p.38). Tarsiers are primitive in having very small brains, but specialized in their extremely long ankle bones (the tarsus) and long hind legs, which enable their huge leaps from stem to stem in the rain forests of island Southeast Asia. Their enormous eyes compensate for them being nocturnal but having no eyeshine, enabling them to hunt for prey at night.

PHILIPPINE TARSIER
Tarsiers are the only living primates that are strictly carnivorous, eating mostly insects, but they also prey on lizards, snakes, birds, and bats. Philippine and Western tarsiers have the largest eyes relative to their body size of all mammals.

Origins of the anthropoids

Anthropoids (monkeys, apes, and humans), like tarsiers, are descended from the omomyids (see p.41), sharing with them shortened faces, and features of the teeth and bony ear. *Algeripithecus* from the Early Eocene of North Africa was thought to be the earliest fossil anthropoid, but new fossil finds in 2009 indicate it may be a strepsirrhine. Undisputed anthropoids date from the Middle Eocene, about 40 million years ago, with fossil jaws and a few skull fragments and foot bones found in Egypt, China, Myanmar, and Thailand. The best known is the tiny *Eosimias*, which weighed only 2–4½oz (60–130 g), and had long, curved canines. A fragmentary cranium from Myanmar has a bony wall at the back of the orbits—a typical anthropoid characteristic.

AEGYPTOPITHECUS
Aegyptopithecus lived in Fayum, Egypt, during the Early Oligocene. It was arboreal, and its teeth indicate it ate fruit.

sagittal crest

bony wall at back of orbit

nuchal crest

long canine

one of eight premolars

FAYUM RESEARCH SITE

The Fayum (Faiyûm) in Egypt is a very important Paleogene fossil primate site, with deposits spanning 3 million years across the Eocene–Oligocene boundary (33.9 million years ago). Several primate lineages have been uncovered at the site, which was first excavated in the early 20th century and, since the 1960s, by an American–Egyptian team led by Elwyn Simons.

New World monkeys

The platyrrhines of South and Central America, known as New World monkeys, all live in forested areas and are highly arboreal. The night monkeys are the only nocturnal group; all other New World monkeys are active by day and have good color vision.

Reaching the New World

At the beginning of the Paleogene, 65 million years ago, South America was a completely isolated continent: the South Atlantic was already half as wide as it is today, and a vast sea separated it from North America (see p.40). Early Paleocene South America was home to monotremes, marsupials, xenarthrans (armadillos, sloths, and anteaters), and unusual hoofed mammals called meridiungulates. The monotremes and meridiungulates are extinct in South America; the marsupials and xenarthrans are still there. There is then a long gap in the fossil record until the Late Paleogene,

during which time the global climate cooled, although South America remained mainly warm and the vegetation tropical, and suddenly platyrrhines appear. How they got there, and from where, is a mystery—perhaps they traveled on rafts of vegetation or via a temporary land bridge across the South Atlantic. The earliest New World monkey, *Branisella*, is known from fossils at a 25–26-million-year-old site in Bolivia. Miocene sites in Chile and Argentina have revealed many monkey fossils.

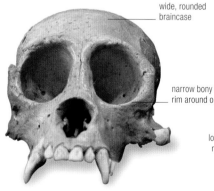

LAND OF PLENTY
Centered around the vast Amazon basin, South America's rain forests contain 20 percent of all the world's plant species, and have supported an extraordinary diversification of New World monkeys.

wide, rounded braincase

narrow bony rim around orbit

CAIPORA
Related to spider monkeys, but twice the size (about 44 lb/20 kg), an almost complete skeleton of *Caipora* was discovered in a Late Quaternary cave in Bahia, Brazil, in 1992.

strong crest for muscle attachments

long, pointed nasal bones

heavy bony rim around orbit

PROTOPITHECUS
Distantly related to *Caipora*, *Protopithecus* was another giant monkey (55 lb/25 kg) from Brazil. Both were probably hunted to extinction by humans.

thick, shaggy fur

prehensile tail used as a fifth limb when climbing through trees

thumb aligned with other fingers

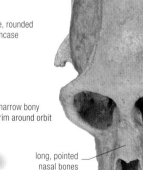

HOWLER MONKEY

Modern monkeys

A monkey is not just a monkey. New World monkeys belong in a group of their own, the platyrrhines, whereas Old World monkeys belong together with apes and humans in a separate group, the catarrhines. They are descended from a common ancestor that lived about 45 million years ago, probably in Africa; the catarrhines remained in Africa and Eurasia, while ancestral platyrrhines made their way to South America. The most conspicuous differences between the two groups are that platyrrhines have a broad nasal septum and the nostrils face to the side, whereas in catarrhines the septum is narrow and the nostrils face forward and down; catarrhines have a more divergent, opposable thumb; and platyrrhines have three premolars (in each half of each jaw), whereas catarrhines have two. The larger platyrrhines have prehensile (grasping) tails, but no catarrhine does. The New World monkeys are split into three families: pitheciids, atelids, and cebids.

flat, receding forehead

RED UAKARI MONKEY

PITHECIIDS
Uakari and saki monkeys live in the Guyanas and the central Amazon region, especially in flooded forest. These small- to medium-sized monkeys feed predominantly on seeds. Their relatives the titi monkeys are more widespread and have a more eclectic diet, feeding mostly on fruit, but also leaves, insects, and small vertebrates.

nostril faces sideways

prehensile tail

BROWN SPIDER MONKEY

BROWN WOOLLY MONKEY

ATELIDS
Atelids are the largest New World monkeys, weighing up to 22 lb (10 kg), and all have prehensile tails. Howler monkeys are the most widely distributed, from southern Mexico to Argentina. Their name comes from their resonant early morning calls. Spider and woolly monkeys and the critically endangered muriqui are restricted to rain forests.

heavy beard, covering the enormously expanded hyoid bone where the voice resonates

long, slender hands, used to probe crevices for insects

CEBIDS
In this diverse family of relatively small monkeys all except the capuchins have long, nonprehensile tails. The marmosets eat tree gum or sap, tamarins feed mostly on fruit, and squirrel monkeys are more insectivorous. The nocturnal night monkeys have less acute color vision than other monkeys, but they do have a good sense of smell.

GOLDEN-HEADED LION TAMARIN

NIGHT MONKEY

GEOGRAPHIC ISOLATION

New techniques of analyzing biodiversity over the past 20 years, especially through DNA sequencing, have revealed many new species of primate. In the Amazon rain forest, isolated populations of howler monkeys and marmosets, for example, have been found to differ in their DNA.

Early apes and Old World monkeys

Hominoids (apes and humans) are best recognized by their specialized backbones and limbs, while Old World monkeys are distinguished by their highly modified molar teeth. The two groups split from their common ancestor about 30 million years ago, during the Oligocene.

smooth brow

face projects forward

braincase same size as living monkeys of similar size

PROCONSUL
These tailless, arboreal quadrupeds lived in East Africa 23–17 million years ago. The genus is related to apes but not ancestral to them. The four known species of *Proconsul* varied in size from about 44–200 lb (20–90 kg).

teeth indicate soft-fruit diet

New opportunities

From the start of the Neogene 23 million years ago, the Earth's climate became gradually cooler and drier, causing deserts and grasslands to spread at the expense of forests. India continued to drift north into Eurasia, closing the Tethys Sea and pushing the Himalayas up into a formidable barrier. Sea levels alternately rose and fell, separating Africa and Eurasia and then connecting them again. Rain forest disappeared from North America and northern Eurasia, but was at times continuous from Africa through southern and southeastern Asia. Primates could spread throughout these areas when forest was continuous, and then evolve and diversify in isolation when the forest split up. The advance of grasslands in Africa may have contributed to the hominoids moving out of the forests and adopting a more upright posture.

African monkeys

The earliest-known fossil Old World monkeys, the primitive victoriapithecid family, lived in Africa about 19–15 million years ago. Their molar teeth were less fully specialized than those of the living cercopithecid family (see opposite), and they probably ate hard fruits or seeds. At first much less numerous than the primitive apes, gradually through the Miocene Old World monkeys became more and more abundant and diverse, later spreading out of Africa into Asia and Europe, while the apes declined, probably because of the competition.

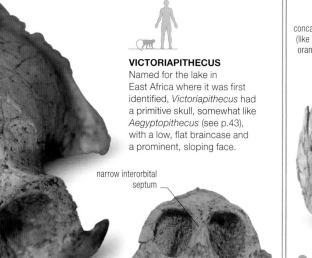

VICTORIAPITHECUS
Named for the lake in East Africa where it was first identified, *Victoriapithecus* had a primitive skull, somewhat like *Aegyptopithecus* (see p.43), with a low, flat braincase and a prominent, sloping face.

narrow interorbital septum

long, grooved canine tooth

The first apes

The first possible ape is the chimpanzee-sized *Kamoyapithecus*, which lived in Africa during the Late Oligocene (about 27–24 million years ago). Because many fossil primates are known only from teeth, and hominoids have more primitive teeth than Old World monkeys, a great many Oligocene and Miocene catarrhines have probably been wrongly classified as "apes" because only their teeth have been found. But there is no doubt that apes were numerous and diverse in the Early Miocene. *Proconsul* (see above) is the best known, but there were many others, mainly from East Africa. The smallest "ape" was *Micropithecus*, with the adult males weighing only 9½ lb (4.3 kg).

elongated, oval orbit

concave face (like modern orangutans)

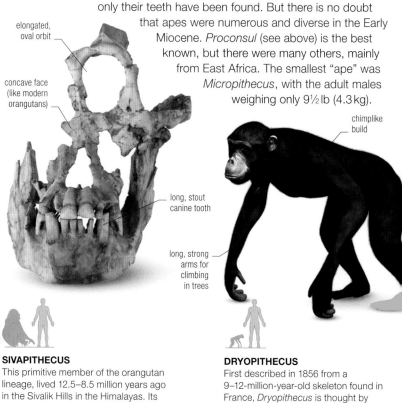

chimplike build

long, stout canine tooth

long, strong arms for climbing in trees

SIVAPITHECUS
This primitive member of the orangutan lineage, lived 12.5–8.5 million years ago in the Sivalik Hills in the Himalayas. Its heavy molars suggest *Sivapithecus* ate tough food such as seeds and hard fruits.

DRYOPITHECUS
First described in 1856 from a 9–12-million-year-old skeleton found in France, *Dryopithecus* is thought by many to be a primitive member of the human-chimpanzee-gorilla group.

Old World monkeys

The cercopithecid family contains at least 140 living species. It is the most diverse and successful family of living primates, and has been so for the past 10 million years. The secret to their success is probably their specialized dentition: they have very hard-wearing molar teeth and long, daggerlike canines. All Old World monkeys are quadrupedal and diurnal, and they range from being highly arboreal to almost entirely terrestrial. Their faces vary from being very short to long and doglike; their tails range from very short to very long; all have opposable thumbs and flattened nails on every digit; and all have hard, bare sitting pads called ischial callosities. Most species show a high degree of sexual dimorphism in either body size (males may weigh three times as much as females), canine size, or color. They live in Africa and Asia, mostly in the tropics, but some range into snowy climates in Japan and the mountains of western China. There are two subfamilies, the cercopithecines and the colobines.

RED COLOBUS

PROBOSCIS MONKEY

COLOBINES
These monkeys have complex stomachs containing bacteria, which ferment the leaves, and often seeds, that make up much of their diet. They do not have cheek pouches.

CERCOPITHECINES
Baboons, macaques, and guenons all belong to the cercopithecines, which have simple stomachs like those of apes (and humans), and tend to be omnivorous; all have deep cheek pouches in which they store food.

ADAPT AND THRIVE

Many Old World monkeys raid crops in rural areas, and in Asia they even live in urban areas. Rhesus monkeys in India often roam villages, towns, and big cities, stealing food. Gray langurs (see right) are much less destructive. They are sacred in Hinduism and their presence is encouraged because they tend to scare away the smaller rhesus monkeys.

NOSTRILS
Unlike New World monkeys, the nostrils in Old World monkeys are separated by a very narrow septum and face forward.

TEETH
The canines are long and sharp, especially in males, and the molars are bilophodont, meaning that they have two transverse ridges.

PLAIN RUMP
The rump has hard sitting pads (ischial callosities), which rest on modified bony bases.

MANDRILL

ALPHA RUMP
In male mandrills, the rump is red and blue, mimicking the color of the face. Both the face and rump become especially bright when a male achieves the dominant alpha status.

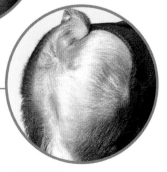

SNOW MONKEYS
The most northerly Old World monkeys live in Honshu, Japan, where there is deep snow in the winter. In some areas, Japanese macaques have learned to keep warm by sitting in hot springs, where they often groom each other.

Modern apes

Chimpanzees, gorillas, and humans form a group of primates often called the "African apes," which many authorities argue is descended from an ancestor that resembles *Dryopithecus* (see p.46); its Asian contemporary, *Sivapithecus*, gave rise to orangutans. Gibbons diverged earlier.

long, curved
fingers

extremely
long arms
attached to
highly flexible
shoulders

What is an ape?

The teeth of apes are less specialized than those of Old World monkeys, but their skeletons are more specialized. Apes have no tail; the caudal vertebrae, which form the tail in most mammals, are reduced, partly fused, and curved forward, forming the coccyx, or tailbone (see p.52). Apes sit or stand upright a good deal. Their lumbar vertebrae are reduced in number and are short and stout to bear the weight of the upper body. They have broad chests, and their shoulder joints are highly mobile, so they can extend their arms out to the sides and above their heads.

Gibbons (all members of the hylobatid family) have extremely long arms, which they use to swing from branch to branch (a type of locomotion called brachiation). They are medium-sized apes, and their diet consists mostly of fruit.

short face with
large orbits

**JUVENILE GIBBON
SKELETON**
The very long arms of gibbons overshadow the fact that they also have long legs; they often walk bipedally along branches or on the ground.

long, curved toes

ischial
callosities
(like Old World
monkeys,
but unlike
great apes)

SIAMANG
Found in the rain forests of Southeast Asia, the siamang is the largest of the gibbons, weighing up to 30 lb (14 kg).

The great apes

Great apes are much larger than gibbons (which are often called lesser apes), more stockily built, and have shorter arms and legs and short thumbs. Orangutans, which are highly arboreal, are restricted to rain forests and today are found only on the islands of Borneo and Sumatra. The much more terrestrial gorillas live in a variety of forest types, and chimpanzees, which are equally at home in the trees and on the ground, can even live in wooded savanna. All, like gibbons, are mainly fruit eaters, but gorillas also eat terrestrial herbaceous vegetation when available seasonally or in habitats where they cannot get fruit, and chimpanzees often also prey on monkeys, small antelopes, or other small mammals.

Chimpanzees live in communities up to 120-strong, in which the males are related, whereas females often leave and join another group. Gorillas live in troops of 10–40, led by a silverback male. Orangutans are solitary, although youngsters will remain with their mother until they are seven or eight years old. Male orangutans and gorillas weigh more than twice as much as the females.

pink facial skin of
juvenile will darken
with age

COMMON CHIMPANZEE
The relaxed posture and facial expression of this young chimp reflect its mood. Like most primates, chimpanzees use their faces and bodies to advertise their intentions.

grasping hands
and feet

BONOBO
A slighter built species of chimpanzee from the tropical forests of the Democratic Republic of Congo, bonobos exhibit different behaviors from those of common chimpanzees, especially the use of sex between individuals of all ages as a way to reduce tensions in the community.

BORNEAN ORANGUTAN
The name orangutan is Malay for "man of the woods." They typically have long reddish brown hair, whereas the other great apes have black or brownish black hair. Most of their waking hours are spent foraging for food.

WESTERN GORILLA
Like all apes, gorillas are completely dependent on their mothers when born and spend the first five months of their lives in constant contact with them. Infants are weaned at three or four years, but the maternal bond remains strong.

Ape anatomy compared

Some of the most important differences between the skeletons of great apes and modern humans are related to their different ways of moving around. While all habitually sit upright, and often also stand upright, which requires a short and sturdy lower spine, humans are the only ones to always move by walking upright on two legs (bipedalism). Gorillas and chimpanzees move by knuckle-walking—a quadrupedal gait in which the fingers are flexed and the weight is supported on the middle finger bones, requiring very strong wrists and hands. When orangutans walk on all fours on the ground, they place their weight on the outside edges of their palms, but much of their lives are spent climbing in trees, and this also requires a highly modified skeleton. The big toe in humans is aligned with the other toes, whereas in gorillas (the next most terrestrial ape), it is fairly divergent, and in chimpanzees, more so. In orangutans, the big toe is very divergent and very short—their feet are modified like hands, to serve their "four-handed" movement through the trees.

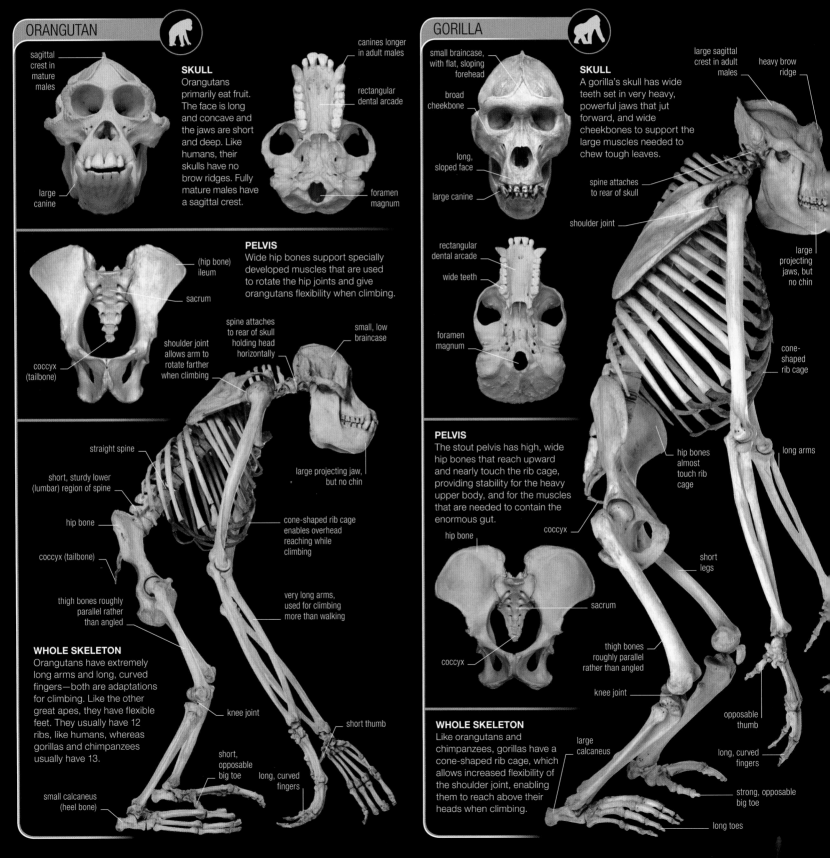

ORANGUTAN

sagittal crest in mature males

large canine

canines longer in adult males

rectangular dental arcade

foramen magnum

SKULL
Orangutans primarily eat fruit. The face is long and concave and the jaws are short and deep. Like humans, their skulls have no brow ridges. Fully mature males have a sagittal crest.

PELVIS
Wide hip bones support specially developed muscles that are used to rotate the hip joints and give orangutans flexibility when climbing.

(hip bone) ileum

sacrum

coccyx (tailbone)

shoulder joint allows arm to rotate farther when climbing

spine attaches to rear of skull holding head horizontally

small, low braincase

large projecting jaw, but no chin

straight spine

short, sturdy lower (lumbar) region of spine

hip bone

coccyx (tailbone)

cone-shaped rib cage enables overhead reaching while climbing

thigh bones roughly parallel rather than angled

very long arms, used for climbing more than walking

WHOLE SKELETON
Orangutans have extremely long arms and long, curved fingers—both are adaptations for climbing. Like the other great apes, they have flexible feet. They usually have 12 ribs, like humans, whereas gorillas and chimpanzees usually have 13.

knee joint

short thumb

short, opposable big toe

long, curved fingers

small calcaneus (heel bone)

GORILLA

small braincase, with flat, sloping forehead

broad cheekbone

long, sloped face

large canine

large sagittal crest in adult males

heavy brow ridge

SKULL
A gorilla's skull has wide teeth set in very heavy, powerful jaws that jut forward, and wide cheekbones to support the large muscles needed to chew tough leaves.

spine attaches to rear of skull

shoulder joint

large projecting jaws, but no chin

rectangular dental arcade

wide teeth

foramen magnum

cone-shaped rib cage

PELVIS
The stout pelvis has high, wide hip bones that reach upward and nearly touch the rib cage, providing stability for the heavy upper body, and for the muscles that are needed to contain the enormous gut.

hip bone

coccyx

hip bones almost touch rib cage

long arms

short legs

sacrum

thigh bones roughly parallel rather than angled

coccyx

knee joint

opposable thumb

WHOLE SKELETON
Like orangutans and chimpanzees, gorillas have a cone-shaped rib cage, which allows increased flexibility of the shoulder joint, enabling them to reach above their heads when climbing.

large calcaneus

long, curved fingers

strong, opposable big toe

long toes

The head in humans is positioned directly above the spine so the foramen magnum (the large hole through which the spinal cord passes into the brain) is positioned much further forward on the skull than it is on the skulls of the great apes. One of the most important characteristics of primates is their intelligence.All primates have large brains compared to their body size, and this is seen to a greater extent in apes. Chimpanzees have a brain size (cranial capacity) of 18.6–29.6 cubic in (305–485 cubic cm); orangutans,

18.4–33.3 cubic in (302–545 cubic cm); and gorillas, 24.6–41 cubic in (403–672 cubic cm). Modern humans, which have the greatest brain to body size ratio of all, have a brain size of 61–122 cubic in (1,000–2,000 cubic cm), and this is reflected in the shape of the braincase. Gorilla and orangutan skulls (and skeletons) show a high degree of difference between the sexes, chimpanzees and humans less so. Some of the other differences in the skulls are related to differences in diet.

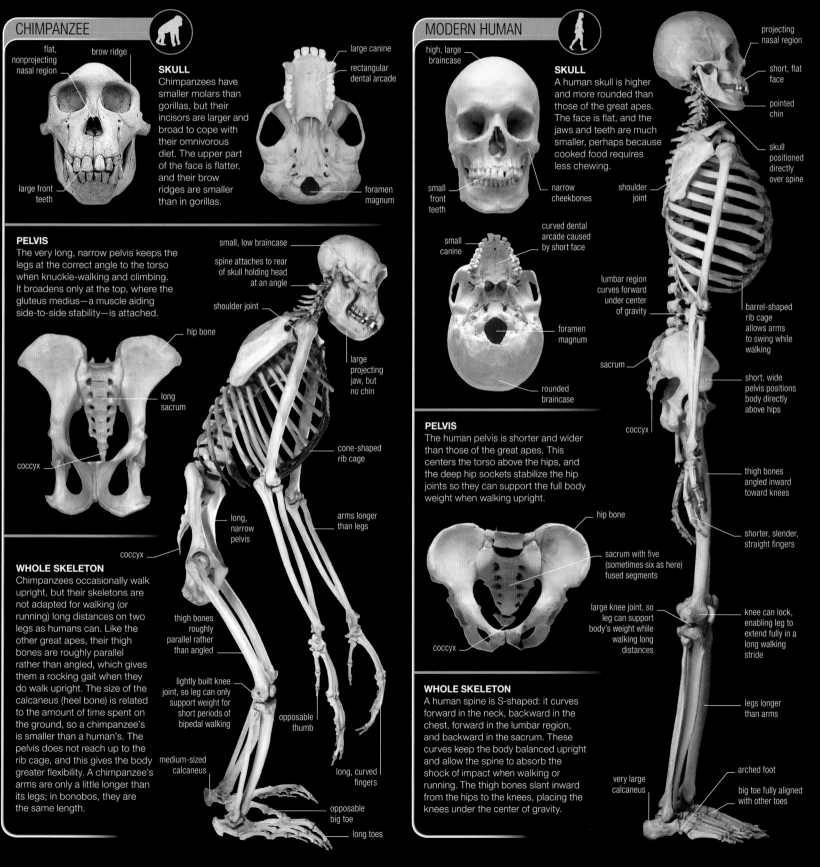

CHIMPANZEE

SKULL
Chimpanzees have smaller molars than gorillas, but their incisors are larger and broad to cope with their omnivorous diet. The upper part of the face is flatter, and their brow ridges are smaller than in gorillas.

- flat, nonprojecting nasal region
- brow ridge
- large front teeth
- large canine
- rectangular dental arcade
- foramen magnum

PELVIS
The very long, narrow pelvis keeps the legs at the correct angle to the torso when knuckle-walking and climbing. It broadens only at the top, where the gluteus medius—a muscle aiding side-to-side stability—is attached.

- hip bone
- long sacrum
- coccyx
- coccyx
- small, low braincase
- spine attaches to rear of skull holding head at an angle
- shoulder joint
- large projecting jaw, but no chin
- cone-shaped rib cage
- long, narrow pelvis
- arms longer than legs

WHOLE SKELETON
Chimpanzees occasionally walk upright, but their skeletons are not adapted for walking (or running) long distances on two legs as humans can. Like the other great apes, their thigh bones are roughly parallel rather than angled, which gives them a rocking gait when they do walk upright. The size of the calcaneus (heel bone) is related to the amount of time spent on the ground, so a chimpanzee's is smaller than a human's. The pelvis does not reach up to the rib cage, and this gives the body greater flexibility. A chimpanzee's arms are only a little longer than its legs; in bonobos, they are the same length.

- thigh bones roughly parallel rather than angled
- lightly built knee joint, so leg can only support weight for short periods of bipedal walking
- medium-sized calcaneus
- opposable thumb
- long, curved fingers
- opposable big toe
- long toes

MODERN HUMAN

SKULL
A human skull is higher and more rounded than those of the great apes. The face is flat, and the jaws and teeth are much smaller, perhaps because cooked food requires less chewing.

- high, large braincase
- small front teeth
- small canine
- narrow cheekbones
- curved dental arcade caused by short face
- foramen magnum
- rounded braincase
- projecting nasal region
- short, flat face
- pointed chin
- skull positioned directly over spine
- shoulder joint
- lumbar region curves forward under center of gravity
- sacrum
- barrel-shaped rib cage allows arms to swing while walking
- short, wide pelvis positions body directly above hips
- coccyx

PELVIS
The human pelvis is shorter and wider than those of the great apes. This centers the torso above the hips, and the deep hip sockets stabilize the hip joints so they can support the full body weight when walking upright.

- hip bone
- coccyx
- sacrum with five (sometimes six as here) fused segments
- large knee joint, so leg can support body's weight while walking long distances
- thigh bones angled inward toward knees
- shorter, slender, straight fingers
- knee can lock, enabling leg to extend fully in a long walking stride

WHOLE SKELETON
A human spine is S-shaped: it curves forward in the neck, backward in the chest, forward in the lumbar region, and backward in the sacrum. These curves keep the body balanced upright and allow the spine to absorb the shock of impact when walking or running. The thigh bones slant inward from the hips to the knees, placing the knees under the center of gravity.

- very large calcaneus
- legs longer than arms
- arched foot
- big toe fully aligned with other toes

Apes and humans

Great apes and humans are grouped as a single family, the hominids, and scientists even consider humans to be a species of ape. Humans belong in the hominin tribe, of which only one species remains today. Genetically, humans are closest to chimpanzees and, reciprocally, chimpanzees are closer to humans than to gorillas.

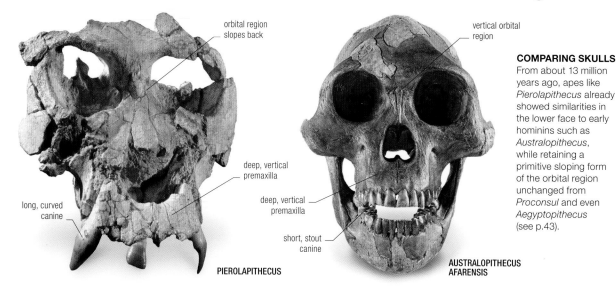

orbital region slopes back

vertical orbital region

long, curved canine

deep, vertical premaxilla

deep, vertical premaxilla

short, stout canine

short, stout canine

PIEROLAPITHECUS

AUSTRALOPITHECUS AFARENSIS

COMPARING SKULLS
From about 13 million years ago, apes like *Pierolapithecus* already showed similarities in the lower face to early hominins such as *Australopithecus*, while retaining a primitive sloping form of the orbital region unchanged from *Proconsul* and even *Aegyptopithecus* (see p.43).

Human or ape?

It is very difficult to draw the line between human and ape in the fossil record. DNA evidence suggests that the common ancestor we share with chimpanzees lived between 10 and 7 million years ago, but finding fossils from this period has become more difficult than it used to be. Scientists have always assumed that the human lineage begins where features associated with walking on two feet (bipedalism; see p.53, 69), short jaws, and short, stout canines are first seen in a species in the fossil record, but this is not clear-cut. It has recently been suggested, controversially, that the common ancestor with chimps and gorillas was already somewhat specialized for bipedalism, but most authorities regard this as unlikely, although it is possible that knuckle-

1. **Leaf rain hat.** Orangutans are especially known for their mechanical flair, such as prying open fruit with a stick. They frequently use large leaves to protect themselves against the rain.

2. **Termite fishing stick.** Chimpanzees have long been known to use grass stems or sticks to extract termites from their mounds. Termiting skills take different forms in different chimpanzee populations.

3. **Leaf sponge.** To extract water to drink from holes in trees, chimpanzees often make "sponges" from leaves. Orangutans have been observed doing this in the wild as well.

4. **Nutcracker.**
Although gorillas use tools much less than chimpanzees or orangutans there are still impressive examples of tool use, such as using stones both as hammer and anvil to crack nuts.

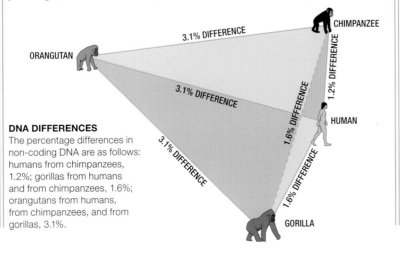

Anything you can do...

All great apes exhibit skills and intelligence well beyond the capabilities of monkeys or even gibbons. They make complex nests for sleeping in at night. They practice strategic planning in their social lives. Wild chimpanzees and orangutans make simple tools and bring them to the sites where they will use them —further evidence of foresight and planning. They also use different tools in sequence; chimpanzees will place a nut on a stone, and use another stone to crack it. One gorilla wading across a pool was seen using a stick to test its depth.

When tested in laboratories, great apes can solve complex psychological problems. They can also recognize themselves in mirrors—like humans, they are self-aware.

LANGUAGE SKILLS
Great apes have been taught simple communication with humans (and with each other). Early programs used sign language, and impressive results have been achieved with bonobos using special symbols.

walking evolved independently in chimps and gorillas. Several evolutionary lines of apes did independently evolve short canines in the Late Miocene. The 7–9-million-year-old *Oreopithecus* from Italy appears to have shown bipedal specializations and had a very short face and small canines, but it was almost certainly not a direct ancestor of our genus, *Homo*.

Ardipithecus ramidus, which lived 4.4 million years ago, was partially bipedal, although still a tree-living ape, and had rather small canines and short jaws. It was only later, during the evolution of *Australopithecus* (from 4.2 million years ago) and then early *Homo* species (from 2.4 million years ago), that the full development of completely human bipedalism, distinctive shapes of teeth, and, finally, brain enlargement is seen.

Close relatives

By how much of their DNA do humans and chimpanzees differ? The percentage is different according to how it is calculated. In non-coding DNA (so-called "junk DNA") it is about 1.2%, but in coding DNA (actual genes) it is, unexpectedly, less, only 0.6%, whereas over their genomes as a whole, because quite large sequences have been deleted and inserted, it is much more, about 5%. Most calculations take the difference in non-coding DNA as the "true" difference. At any rate, humans and chimpanzees are very closely related, and separated between 10 and 7 million years ago, whereas the ancestors of gorillas separated a little over 10 million years ago, and those of orangutans about 15 million years ago.

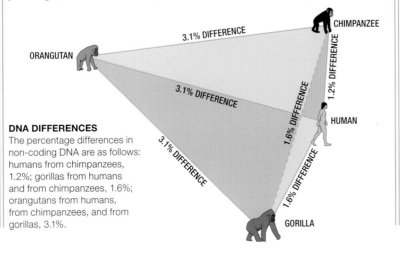

DNA DIFFERENCES
The percentage differences in non-coding DNA are as follows: humans from chimpanzees, 1.2%; gorillas from humans and from chimpanzees, 1.6%; orangutans from humans, from chimpanzees, and from gorillas, 3.1%.

ORANGUTAN
3.1% DIFFERENCE
CHIMPANZEE
3.1% DIFFERENCE
1.2% DIFFERENCE
HUMAN
1.6% DIFFERENCE
3.1% DIFFERENCE
1.6% DIFFERENCE
GORILLA

JANE **GOODALL**

British primatologist Jane Goodall (1934–) began her study of chimpanzees in Gombe National Park, Tanzania, in 1960. She was the first to observe toolmaking by wild chimpanzees. She documented chimpanzee society, their behavioral flexibility, their emotional life, their cooperative hunting, and their wars against neighboring communities.

HOMININS

The term "hominin" describes living humans and all other species comprising the lineage that diverged from that of chimpanzees between 10 and 7 million years ago. The hominin fossil record is currently made up of the 24 species profiled here. Only some of them are our ancestors, and many became extinct without giving rise to new species. Although the relationships between these hominins are complex and often unresolved, various aspects of their fossil remains have enabled us to place them in groups or genera, including *Homo*, to which our own species, *Homo sapiens*, belongs.

Human evolution

The fossil record for human evolution stretches back at least 7 million years. Early sites are now known from eastern, central, and southern parts of Africa, and later sites have been found across many parts of Europe and Asia. After nearly 150 years of scientific investigation, there is strong evidence for the general pattern of human evolution.

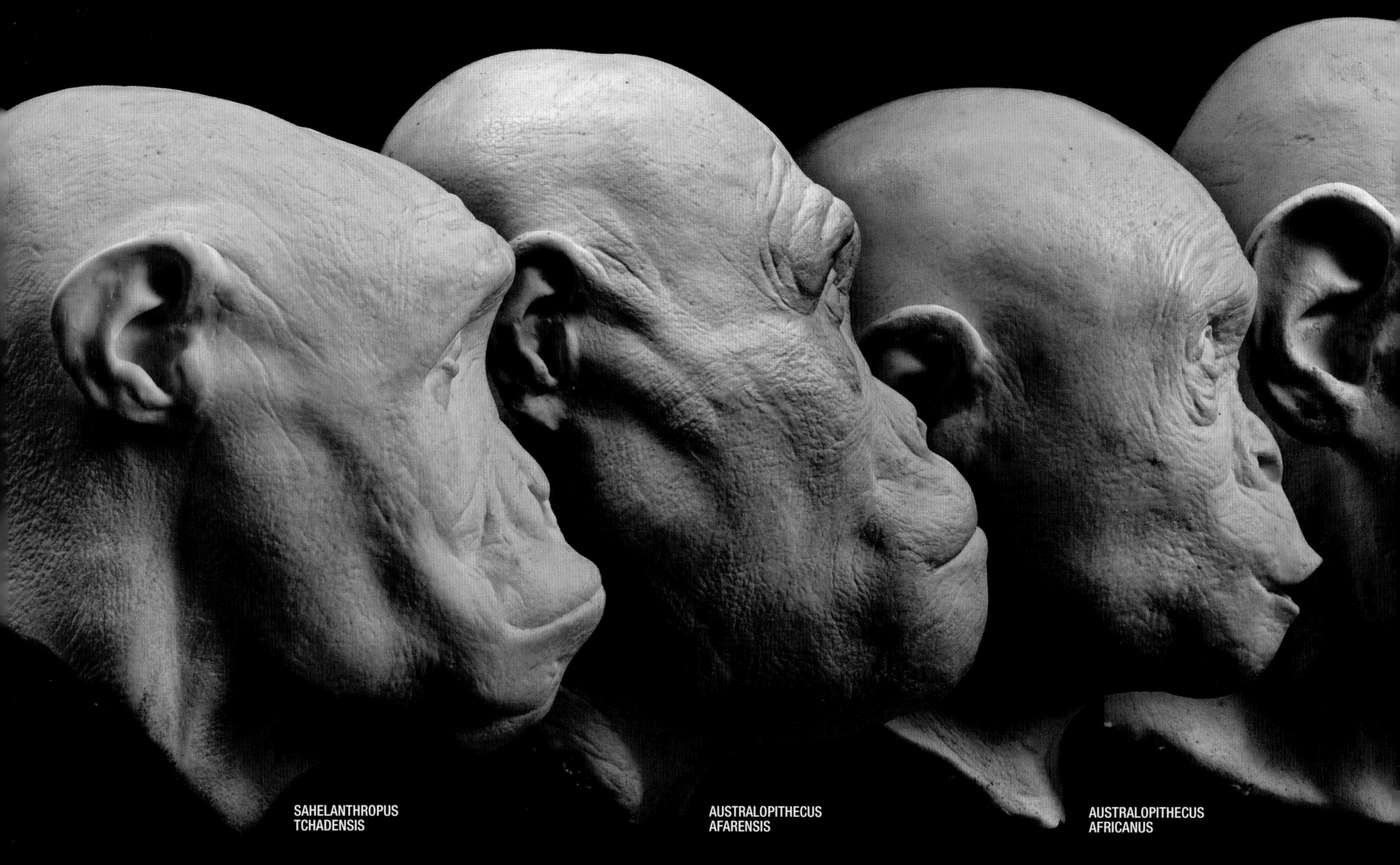

SAHELANTHROPUS TCHADENSIS

AUSTRALOPITHECUS AFARENSIS

AUSTRALOPITHECUS AFRICANUS

Becoming human

Humans have much in common, anatomically and behaviorally, with other primates. This close biological kinship is also indicated by DNA evidence (see p.36), which shows that our closest living relatives are bonobos and chimpanzees. There is only about 1 percent difference between the chimpanzee genome and our own (see p.55), suggesting that we share a common ancestor. Genetic mutations occur at similar rates across different lineages, and accumulated change can be used to estimate the date of species divergence. According to this "molecular clock," the last ancestor we shared with chimpanzees probably lived between 10 and 7 MYA. One of the challenges in paleoanthropology is working out what this common ancestor looked like, how it behaved, and what selection pressures resulted in the appearance of the first hominin

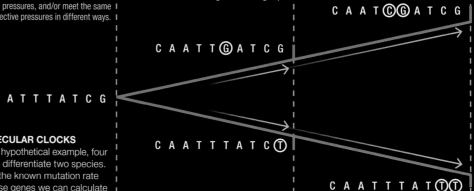

50 MILLION YEARS AGO
Members of the same species have the same genetic makeup. But over time, two different groups within the species face different selective pressures, and/or meet the same selective pressures in different ways.

25 MILLION YEARS AGO
The groups mix less and less; there is less opportunity or willingness to interbreed, and mutations (the circled genes) begin to distinguish the two groups.

PRESENT DAY
Ultimately, sufficient mutations build up in both groups' genomes that they are no longer able to interbreed, so they become distinct species.

C A A T T ⓒⒼ A T C G

C A A T T Ⓖ A T C G

C A A T T T A T C G

MOLECULAR CLOCKS
In this hypothetical example, four genes differentiate two species. From the known mutation rate of those genes we can calculate back for a common ancestor

C A A T T T A T C Ⓣ

C A A T T T A T ⓉⓉ

"Human" species

Many biologists observing living animals define a species as a group of interbreeding natural populations that are reproductively isolated from other such groups. This is the "biological species" concept. Paleoanthropologists cannot observe the behavior of extinct hominin species, so they have to rely on the fossils themselves for clues about their distinctiveness. Variation in the morphology (shape) of bones is used to group fossils into populations, and if these populations are distinctive enough, a new species is proposed. It is even possible that the definition of a species may rely on a single very unusual fossil specimen. As new finds are made and the understanding of the fossil record grows, the names of hominin species may be redefined, grouped, or eliminated.

AFRICAN ANCESTORS
This lineup shows some of the African members of the human evolutionary tree, spanning back 7 million years. While it appears linear here, the map of human evolution looks more like a tree, with several branches, for example *Homo habilis* and *Homo ergaster*, living at the same time.

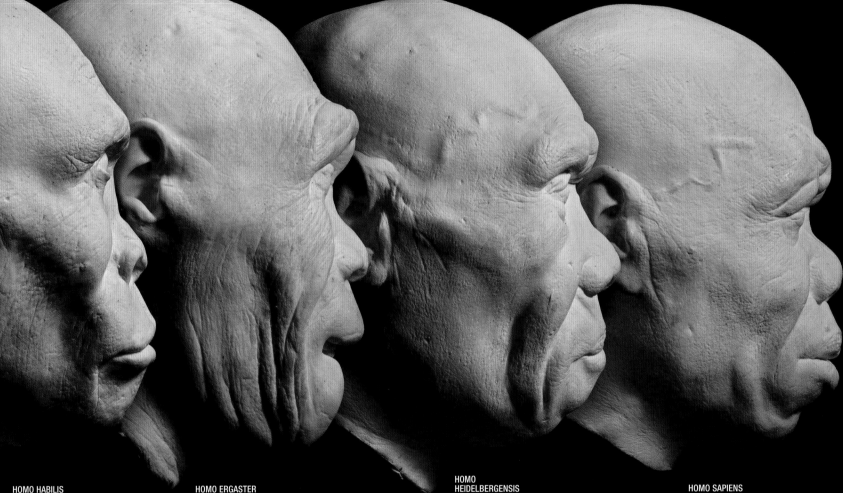

HOMO HABILIS

HOMO ERGASTER

HOMO HEIDELBERGENSIS

HOMO SAPIENS

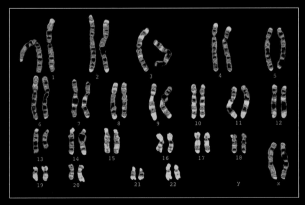

DNA KARYOTYPE
The human genome contains 46 chromosomes, while our close relatives the great apes have 48. In the human lineage, two chromosomes inherited from the ancestors we shared with the other apes have fused together to form the human chromosome 2.

Genetics

The science of genetics reveals the mechanisms at the heart of evolutionary processes. Traits are encoded in genes—hereditary units of DNA in the chromosomes of our cells. Processes such as natural selection and mutation can change the genetic profile of populations and produce new species. Genetic studies show that all living humans are closely related and suggest we shared a common ancestor who lived about 200,000 years ago. Research has also revealed DNA evidence of modern humans interbreeding with Neanderthals and other archaic species.

MITOCHONDRIAL EVE
We inherit mitochondrial DNA (mtDNA) only from our mothers. By comparing differences between mtDNA lineages, and estimating the rate at which they appeared, it is possible to trace the lineages back to a common female ancestor, known as "Mitochondrial Eve."

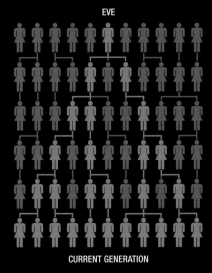

EVE

CURRENT GENERATION

Family tree

Genetic and fossil evidence suggests that the first hominins appeared in Africa between 8 and 6MYA, with many species appearing after this time. Their genetic relationships probably form a complex web, and until relatively recently several hominin species existed at any one time.

Trends through time

Several specimens, such as *Sahelanthropus tchadensis* and *Orrorin tugenensis*, suggest that the earliest hominin species were modest in size, had brains no larger than those of modern apes, and exhibited a unique suite of physical characteristics that allowed both upright walking and climbing. Over time, and in the context of changing selection pressures, populations appeared with new characteristics, and a number of evolutionary trends can be identified. Powerful jaws and large back teeth, ideal for chewing tough or fibrous foods, appear in some species. Large brains relative to body size and smaller jaws and teeth appear in others. Bipedalism became the dominant mode of locomotion, and all the later hominins were characterized by their use of stone-tool technology.

KEY TO GENUS

	SAHELANTHROPUS
	ORRORIN
	ARDIPITHECUS
	KENYANTHROPUS
	PARANTHROPUS
	AUSTRALOPITHECUS
	HOMO

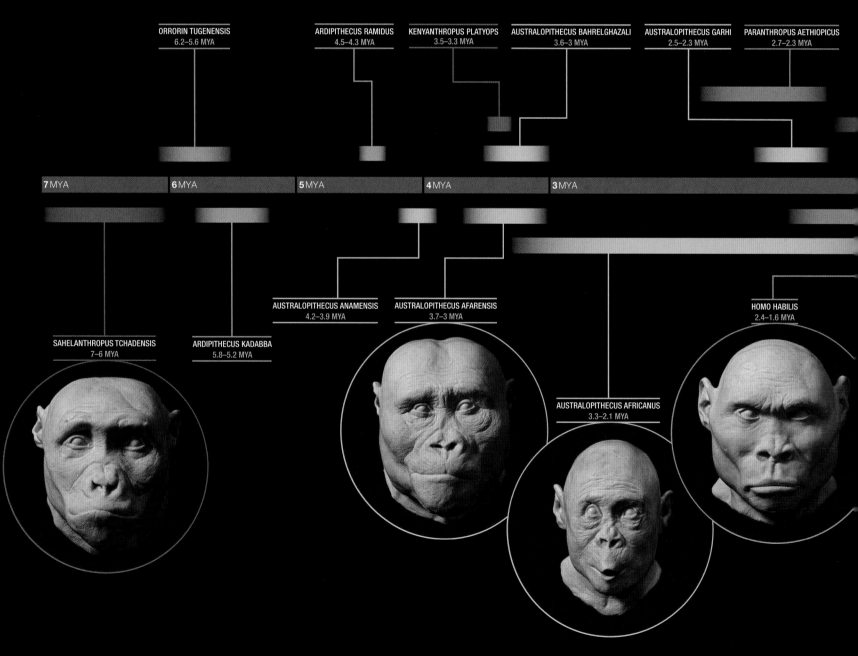

ORRORIN TUGENENSIS
6.2–5.6 MYA

ARDIPITHECUS RAMIDUS
4.5–4.3 MYA

KENYANTHROPUS PLATYOPS
3.5–3.3 MYA

AUSTRALOPITHECUS BAHRELGHAZALI
3.6–3 MYA

AUSTRALOPITHECUS GARHI
2.5–2.3 MYA

PARANTHROPUS AETHIOPICUS
2.7–2.3 MYA

7 MYA   6 MYA   5 MYA   4 MYA   3 MYA

AUSTRALOPITHECUS ANAMENSIS
4.2–3.9 MYA

AUSTRALOPITHECUS AFARENSIS
3.7–3 MYA

HOMO HABILIS
2.4–1.6 MYA

SAHELANTHROPUS TCHADENSIS
7–6 MYA

ARDIPITHECUS KADABBA
5.8–5.2 MYA

AUSTRALOPITHECUS AFRICANUS
3.3–2.1 MYA

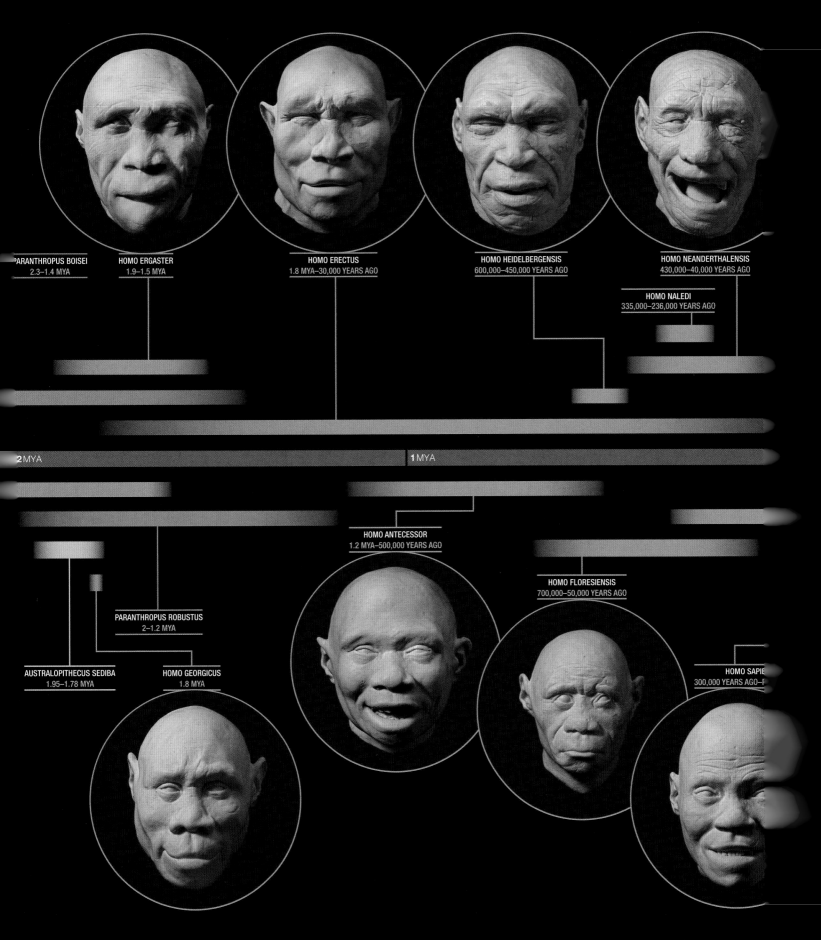

PARANTHROPUS BOISEI
2.3–1.4 MYA

HOMO ERGASTER
1.9–1.5 MYA

HOMO ERECTUS
1.8 MYA–30,000 YEARS AGO

HOMO HEIDELBERGENSIS
600,000–450,000 YEARS AGO

HOMO NEANDERTHALENSIS
430,000–40,000 YEARS AGO

HOMO NALEDI
335,000–236,000 YEARS AGO

2 MYA

1 MYA

HOMO ANTECESSOR
1.2 MYA–500,000 YEARS AGO

HOMO FLORESIENSIS
700,000–50,000 YEARS AGO

PARANTHROPUS ROBUSTUS
2–1.2 MYA

AUSTRALOPITHECUS SEDIBA
1.95–1.78 MYA

HOMO GEORGICUS
1.8 MYA

HOMO SAPIE
300,000 YEARS AGO–

Sahelanthropus
tchadensis

Sahelanthropus lived around the time of the last common ancestor of humans and other apes, but its position on the human family tree is far from certain.

SAHELANTHROPUS TCHADENSIS

> **NAME MEANING** "The Sahel man from Chad"

> **AGE** 7–6 mya

> **HEIGHT** Unknown

> **BRAIN SIZE** 19–23 cu in (320–380 cc)

> **LOCALITY** Toros-Menalla, Chad

> **FOSSIL RECORD** Single skull, fragments of jaw and teeth

Sahelanthropus tchadensis, described in 2002, was found in Chad—far away from the East African Rift Valley and South Africa, where the search for early hominins had previously been concentrated.

Discovery

In 2001 the Mission Paléoanthropologique Franco-Tchadienne, a joint venture between universities and research institutes in France and Chad, began to recover fossils from the Toros-Menalla area of Chad's Djurab Desert. There are no absolute dates for these finds, but comparisons with animal fossils from other sites suggest that the material is 7–6 million years old. The fossils, which represent nine individuals, include a relatively complete skull, four jawbone fragments, and a few teeth. No fossils of other body parts are known, making direct comparison with early hominin skeletons difficult. Some of the cranial characteristics are similar to those of Miocene apes, and others more similar to later hominins. However, the fossil finds are separated from these groups by both time and geography, so the new genus seems justified.

FOSSIL SKULL IN SITU
The single *S. tchadensis* skull discovered, known as Toumai ("Hope of Life") in the local language of Chad, was found exposed on the surface in loose sand. This made exact dating difficult, since it may have been moved or even reburied in the recent past.

Physical features

Sahelanthropus has several features in common with later hominins such as *Kenyanthropus* and *Homo*. The canines are relatively small and show wear at their tips, and the tooth enamel is thicker than is seen in apes. The face is also quite flat compared with apes, and the brow ridge is massive and continuous above the orbits. In many other respects, however, *Sahelanthropus* does resemble living and extinct apes. The cranial capacity is small, and the shape of the rear part of the skull is a truncated triangle. It has been argued that because *Sahelanthropus* presents this mosaic of primitive and recently evolved characteristics, it probably should be placed on the human family tree close to the last common ancestor of humans and chimpanzees.

VIRTUAL SKULL RECONSTRUCTION

A computed tomography (CT) scan was used to make a digital version of the *Sahelanthropus* skull. The elongated skull has a horizontal base and a short, vertical face, as in later hominins. The opening for the spinal cord is oval, downward pointing, and positioned toward the front of the skull, suggesting that the head was balanced on an erect spine. These features are shared with later bipedal hominins, such as *Australopithecus africanus* and *Homo sapiens*, but not other apes.

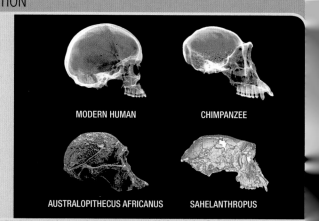

MODERN HUMAN   CHIMPANZEE

AUSTRALOPITHECUS AFRICANUS   SAHELANTHROPUS

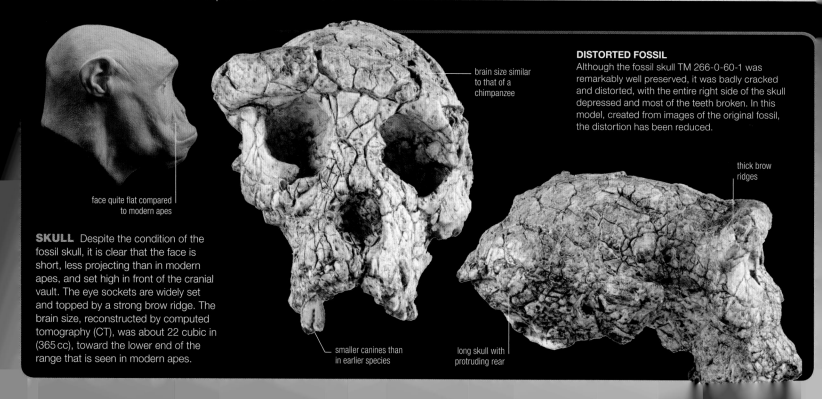

face quite flat compared to modern apes

brain size similar to that of a chimpanzee

DISTORTED FOSSIL
Although the fossil skull TM 266-0-60-1 was remarkably well preserved, it was badly cracked and distorted, with the entire right side of the skull depressed and most of the teeth broken. In this model, created from images of the original fossil, the distortion has been reduced.

thick brow ridges

SKULL Despite the condition of the fossil skull, it is clear that the face is short, less projecting than in modern apes, and set high in front of the cranial vault. The eye sockets are widely set and topped by a strong brow ridge. The brain size, reconstructed by computed tomography (CT), was about 22 cubic in (365 cc), toward the lower end of the range that is seen in modern apes.

smaller canines than in earlier species

long skull with protruding rear

Paleoanthropologists are trying to unravel the role of climate change in human evolution. In sub-Saharan Africa, geological evidence from the last 10 million years suggests a trend toward cooling temperatures, while the uplift of the East African Rift System led to increased dryness. These factors probably caused a shift from woodland to grassland. Superimposed on such long-term trends was extreme climate variability, with with East African lakes fluctuating between high and low water levels.

WOODLAND
African apes still live in forest and woodland habitats, suggesting that they were important in early hominin evolution. Bipedal, ground-living primates like the early hominins probably lived in open, seasonal woodlands rather than in dense tropical forests. Here food resources are less predictable, and exploiting them successfully requires flexibility in diet and in foraging behavior.

GRASSLAND
The African savanna and other grassland habitats are more open than woodlands. They tend to support a larger number and greater diversity of terrestrial mammal and plant species. Many food resources, such as nuts, underground tubers, and bone marrow scavenged from animal carcasses, could only be accessed by hominins using tools.

The Savanna Hypothesis

There are several hypotheses linking African climate change to hominin evolution. The long-held savanna hypothesis states that increasing aridity and the spread of savanna habitats drove the evolution of hominin characteristics such as bipedalism and increased brain size. Evidence for these trends comes from carbon-isotope data in soils (see p.10), faunal remains, ocean sediments, and an apparent correlation in timings of climate change and the appearance of new hominin species. Upright walking, it is suggested, was the most efficient way of moving over open grasslands, with the added benefit that it freed the hands for carrying items or for making tools. However, recent evidence is challenging this view, suggesting that the relationship between climate change and hominin evolution is far more complex.

Pulsed Climate Variability Hypothesis

If key hominin characteristics evolved in response to the selection pressures of a savanna habitat, it is unlikely that hominin fossils with these characteristics would be found in any other habitat type. However, *Ardipithecus* remains have now been found in woodland habitats, and *Australopithecus anamensis* fossils found in association with mixed grassland and gallery forest environments. This suggests that perhaps the hominin skeleton was very adaptable and allowed individuals to flourish in a range of habitats, whether wet or dry, open or wooded. Recent research has shown that Africa experienced extreme climate variability, which in turn would have caused significant variability in vegetation and water availability. The "Variability Selection Hypothesis," developed by Rick Potts, suggests that large climate fluctuations (as opposed to gradual climate change) cause

habitat-specific traits to be replaced by adaptations that are responsive to rapidly changing environments. These ideas have been refined still further by Mark Maslin and Martin Trauth in the "Pulsed Climate Variability Hypothesis," which suggests that the transitional phases between wet and dry periods suffer very extreme climate variability over short timescales, and that it is these pressures that could have had a marked effect on the appearance of new species.

HOMININS AND EXTREME CLIMATE PERIODS
The emergence of many new hominin species seems to have coincided with periods of highly variable climate. It may be that the physical and behavioural flexibility of hominin species evolved so that they could cope in a range of environmental conditions.

HOMININ GENERA

	HOMO
	AUSTRALOPITHECUS
	PARANTHROPUS
	KENYANTHROPUS
	ARDIPITHECUS

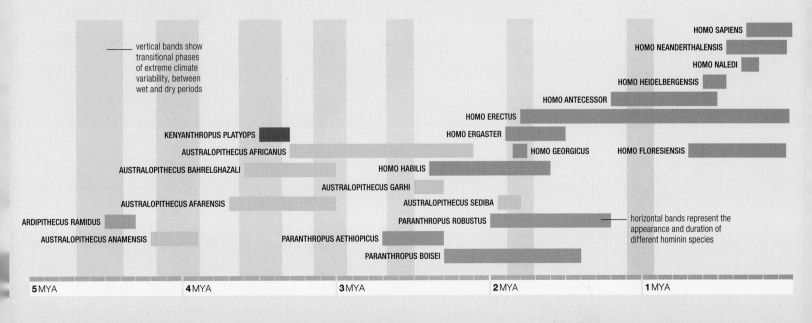

vertical bands show transitional phases of extreme climate variability, between wet and dry periods

HOMO SAPIENS
HOMO NEANDERTHALENSIS
HOMO NALEDI
HOMO HEIDELBERGENSIS
HOMO ANTECESSOR
HOMO ERECTUS
KENYANTHROPUS PLATYOPS
HOMO ERGASTER
AUSTRALOPITHECUS AFRICANUS
HOMO GEORGICUS
HOMO FLORESIENSIS
AUSTRALOPITHECUS BAHRELGHAZALI
HOMO HABILIS
AUSTRALOPITHECUS GARHI
AUSTRALOPITHECUS AFARENSIS
AUSTRALOPITHECUS SEDIBA
ARDIPITHECUS RAMIDUS
PARANTHROPUS ROBUSTUS
AUSTRALOPITHECUS ANAMENSIS
PARANTHROPUS AETHIOPICUS
PARANTHROPUS BOISEI

horizontal bands represent the appearance and duration of different hominin species

5MYA **4**MYA **3**MYA **2**MYA **1**MYA

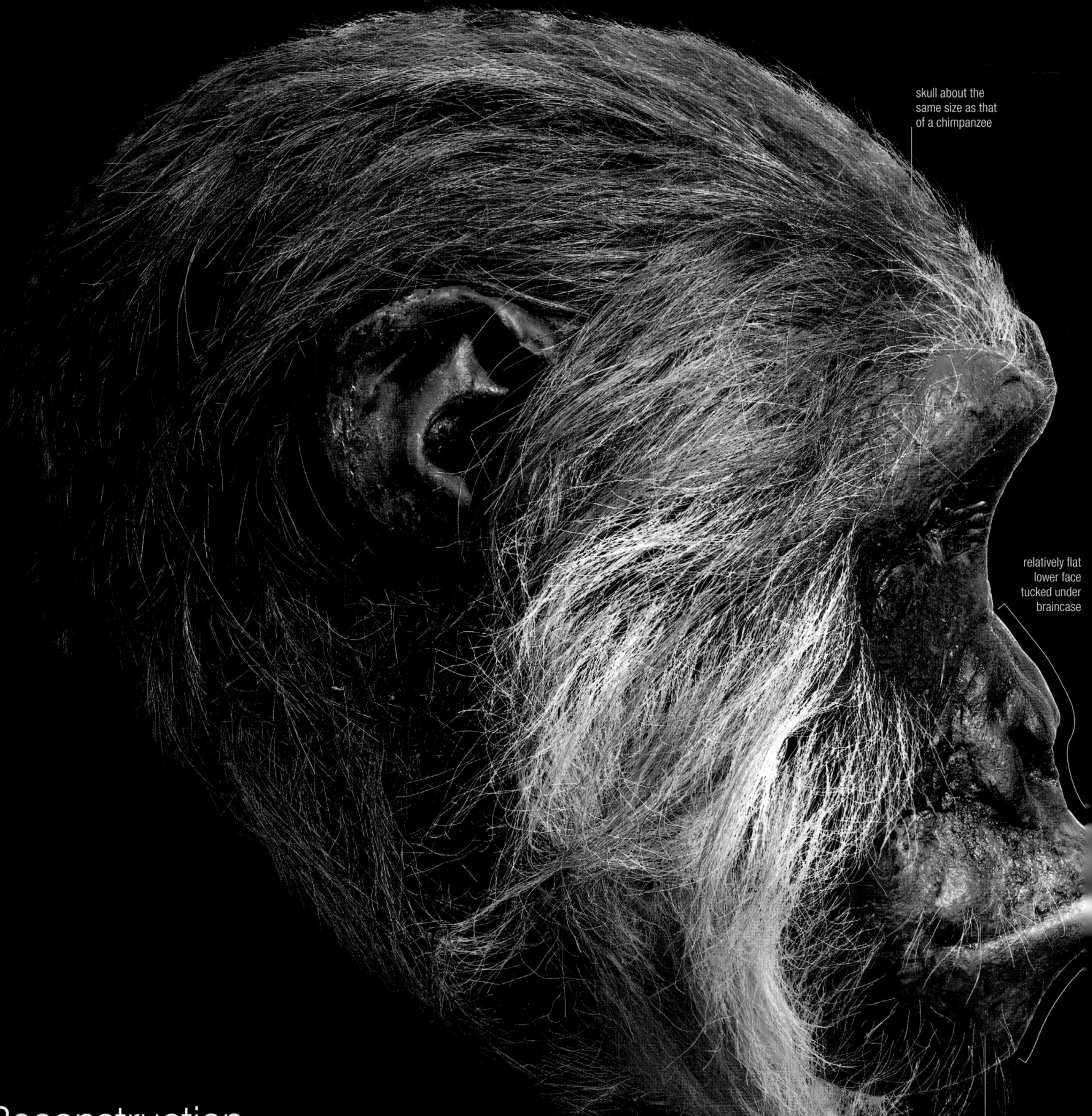

skull about the
same size as that
of a chimpanzee

relatively flat
lower face
tucked under
braincase

relatively small
jaw does not
protrude as far as
in other primates

Reconstruction

This representation of *Sahelanthropus tchadensis* is based on measurements
taken from the only skull yet found of this hominin species. Nicknamed "Toumaï"
("hope of life" in the local Dazaga language of Chad), the skull has a mixture
of primitive, apelike features and others characteristic of later hominins. It
was partly crushed and distorted, and some of its details had been eroded
by wind-blown sand because it had lain exposed in the open. This presented
a challenge to the modelers attempting to reconstruct Toumaï's possible
appearance, necessitating a considerable amount of educated guesswork

eyes, like those of other primates, probably did not have visible whites

typical ape nose is wide and flat, unlike human noses

dark facial skin protects against UV radiation from the sun, and gets darker with age and exposure

heavy ridge of bone over eyes

face surprisingly modern for such an early species

head probably positioned more squarely over an erect spine, compared with knuckle-walking apes, such as gorillas and chimpanzees

Orrorin tugenensis

A contender for the title of the earliest bipedal hominin, *Orrorin tugenensis* may have inhabited ancient lakeside woodlands and wet grasslands.

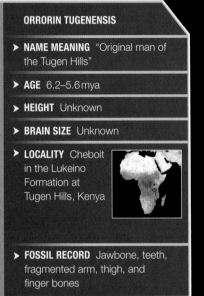

ORRORIN TUGENENSIS

- **NAME MEANING** "Original man of the Tugen Hills"
- **AGE** 6.2–5.6 mya
- **HEIGHT** Unknown
- **BRAIN SIZE** Unknown
- **LOCALITY** Cheboit in the Lukeino Formation at Tugen Hills, Kenya
- **FOSSIL RECORD** Jawbone, teeth, fragmented arm, thigh, and finger bones

Discovery

In 1974, a fossil molar tooth was unearthed by palaeontologist Martin Pickford from late Miocene deposits at Cheboit, Kenya. With low cusps on the chewing surface and thick enamel, it did not fit into any known species. Pickford named it *Orrorin tugenensis* the following year, but it was nearly 30 years until the next major find. A French–Kenyan team led by Pickford and French palaeontologist Brigitte Senut returned to the region and in 2001 announced the discovery of teeth and fragments of the arms and legs of several individuals. This species may have been one of the earliest members of the human lineage to walk upright.

Physical features

The shaft of a human femur (thigh bone) is a very strong cylinder, adapted to the heavy loading of bipedal walking. In *Orrorin tugenensis*, the femur has particularly thick bone in the upper part of the shaft—evidence for bipedalism. The shape of the upper arm bone and the slightly curved finger, however, indicates that the upper limbs were weight bearing too. The pointed canine tooth, low cusped molars, and large upper front tooth may reflect a diet of fruit and seeds. Taken together, this could imply a mixed arboreal and terrestrial lifestyle.

parts of three separate femurs were discovered, although only one was well-preserved enough to prove *Orrorin* walked on two legs

ORRORIN THIGH BONE
Several features of *Orrorin*'s femur suggest that it walked on two legs. The round shape of the ball joint that fits into the hip socket, its orientation to the shaft, and the position of marks from hip muscles on the bone are like those of bipeds.

Ardipithecus kadabba

Notable for its large, projecting canine teeth, *Ardipithecus kadabba* may be related to later hominin species such as *Australopithecus anamensis* (see p.75) and *Australopithecus afarensis* (see pp.78–79).

ARDIPITHECUS KADABBA

- **NAME MEANING** From the Afar word for "basal family ancestor"
- **AGE** 5.8–5.2 mya
- **HEIGHT** Unknown
- **BRAIN SIZE** Unknown
- **LOCALITY** Middle Awash, Ethiopia
- **FOSSIL RECORD** Fragments of jaw, arm, hand, foot bone, and collarbone

Discovery

In 2004, anthropologists Tim White (USA), Gen Suwa (Japan), and Berhane Asfaw (Ethiopia) renamed a small collection of fossil fragments from the Middle Awash River Valley, Ethiopia, as *Ardipithecus kadabba*. Excavated over the previous decade, the fossils were first thought to be an australopithecine species, and then a subspecies of *Ardipithecus ramidus*. Now the material is seen as a species in its own right, largely on the basis of dental anatomy and geological date. The fossils were found with remains of extinct animals such as the four-tusked elephant *Deinotheirium* and the three-toed horse *Hipparion*, as well as wetland and woodland species still alive today.

AWASH LANDSCAPE
Ardipithecus kadabba's habitat was a mix of woodland and grassland, with springs, swamps, and small lakes—not quite as arid as the Awash landscape appears today.

Physical features

The teeth of *Ardipithecus kadabba* share some characteristics with those of great apes. The canines are large and project past the chewing plane of the molars. They are also worn along the length of the tooth—in modern apes, this is caused by the teeth interlocking. However, the molar enamel is thicker than that of chimpanzees, but thinner than in later hominins. It is likely that this species enjoyed a diet of fruit and soft leaves.

Some fragmented limb bones have been ascribed to *A. kadabba*, but they remain controversial. The orientation and shape of the toe may suggest a grasping foot, and the dimensions of the arm may indicate a body size like australopithecines, but until a more complete *A. kadabba* skeleton is found, few conclusions can be drawn.

BIPEDALISM
ANATOMICAL FEATURES

Upright walking on two legs, or bipedalism, is one of the most distinctive human features. Other primates may stand on their hind limbs to reach food, or briefly walk bipedally to cross open spaces, but they swiftly return to quadrupedalism or climbing. Bipedalism gave our ancestors many advantages: it was an efficient way of traveling long distances; it helped maintain a moderate body temperature by minimizing exposure to the sun; it raised the line-of-sight for spotting predators; and it freed the hands for using tools. The adaptations that allow us to walk bipedally can be found all over our skeletons.

○ GORILLA   ○ HUMAN

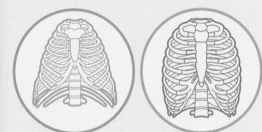

RIB CAGE
The human rib cage is barrel shaped so that the torso can flex and the arms swing freely, which aids balance while walking bipedally. Gorillas have a cone-shaped rib cage to accommodate a large gut below and permit a wide range of motion at the shoulder, enabling them to reach above their head when climbing.

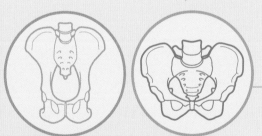

PELVIS
The human pelvis is shorter and wider than that of other primates, with the base of the spine close to the large hip-joint socket. This shape supports the upper body, enables the pelvis to tilt, and maintains balance in upright walking.

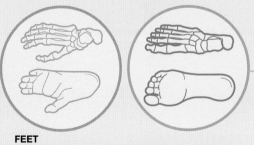

FEET
The feet of most primates are adapted to grasping and quadrupedal movement, with divergent big toes and flat soles. In humans, large big toes aligned with the other toes, well-developed arches, and wide heels allow the foot to push off with the toes and absorb the forces of walking.

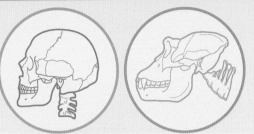

FORAMEN MAGNUM
The spinal cord passes through a large opening in the skull called the foramen magnum. This is at the rear of the skull in quadrupeds, but in bipeds the skull sits on a vertical spine, so the foramen magnum is tucked under the base of the skull.

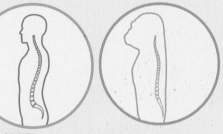

SPINE
To accommodate our upright stance, the human spine is more curved in the neck and lower back than that of a gorilla. This S-shape brings the body's center of gravity close to the midline above the feet and allows the spine to flex during walking.

KNUCKLE-WALKING GORILLA
Our closest modern relatives, the chimpanzees and gorillas, are "knuckle-walkers," resting on their knuckles, rather than their palms or fingers, while moving on the ground.

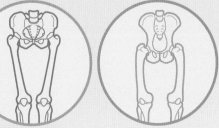

THIGH BONE
The human femur (thigh bone) transfers the body's weight from the pelvis to the feet through the knees, so it has large joint surfaces and a shaft angled toward the body's midline and center of gravity. In contrast, most other primates have shorter, less angled femurs with smaller joints.

Ardipithecus ramidus

Ardipithecus ramidus is known from the remains of many individuals, representing almost the entire skeleton, and it provides an insight into the appearance of bipedalism and habitat selection in early hominin evolution.

Discovery

In 1992, a single hominin molar tooth was discovered by the Japanese anthropologist Gen Suwa in the Aramis locality 1 site of the Middle Awash River Valley in Ethiopia. This locality was known to contain a rich paleontological record, including fragmented remains of woodland plants and animals, but this was the first hominin evidence found there. The specimen remained unclassified until 1993, when a research team led by American anthropologist Tim White found the remains of at least 17 individuals, mostly represented by teeth. Together, these were used to describe a new species: *Australopithecus ramidus*. In 1995, the species genus was changed to *Ardipithecus*. Today, the species is represented by more than 100 specimens, including the remarkably complete skeleton ARA-VP-6/500. Many of these fossils come from contexts that can be confidently dated, and the species is thought to have lived between about 4.5 and 4.3 million years ago.

GIDAY WOLDEGABRIEL
The Ethiopian-born geologist Giday WoldeGabriel has been involved in the discovery of many significant fossil specimens, especially during his involvement with the Middle Awash Project in East Africa.

Physical features

The partial remains of many individuals have been discovered and described—unusual for such an ancient species—and fossils of almost every part of the skeleton have been found. Analysis of these specimens has shown that *Ardipithecus ramidus* was of modest stature and probably presented very little sexual dimorphism (size difference between the sexes). Many features of the arms, hands, and feet suggest that they employed a mixed pattern of movement, partially bipedal and partially arboreal. The dentition suggests a broad diet, and the brain is small compared to that of later hominins.

> **HEIGHT** Female: 3ft 11in (1.20m)

> **WEIGHT** Female: 110lb (50kg)

> **BRAIN SIZE** 18–23 cubic in (300–370 cubic cm)

"ARDI" – THE OLDEST HOMININ SKELETON
In November 1994, Ethiopian paleoanthropologist Yohannes Haile-Selassie collected two hominin hand bones from the surface of an exposed clay deposit at Aramis. Soon finger bones, then a thigh bone and a lower leg were revealed, and by the end of 1995 more than 100 more fragments from the same skeleton had been found. The remains were poorly fossilized and fragile, and it took the research team several years of meticulous laboratory work to stabilize them. The partial skeleton ARA-VP-6/500, known as "Ardi," includes a fragmented but quite complete skull, remains of both arms and hands, parts of the pelvis, leg, and both feet. It is identified as female because it has smaller canines, a smaller cranial capacity, and a more lightly built face than other specimens.

ARDIPITHECUS RAMIDUS

> **NAME MEANING** "Root of the ground apes"

> **LOCALITY** Aramis in Middle Awash, and Goma (near Hadar), Ethiopia

> **AGE** *4.5–4.3 mya*
Dated through absolute dating of volcanic ash layers above and below the fossils

8 MYA PRESENT 7 MYA 1 MYA 6 MYA 2 MYA 5 MYA 3 MYA 4 MYA

> **FOSSIL RECORD** One nearly complete skeleton; various skull, jaw, teeth, and arm fragments

ARDI'S ENVIRONMENT

The hominin-bearing deposits of the Middle Awash are rich in evidence of *Ardipithecus ramidus*'s local environment. Fossilized wood fragments, seeds, and the remains of mammals such as antelopes and monkeys all indicate an ancient woodland or forest habitat at a time when drought was rare. Biomolecular studies of the teeth of several *A. ramidus* individuals have suggested that their diet was dominated by the products of shrubs and trees, which also points to a preference for woodland habitats rather than the open savanna and grassland so often associated with later hominins.

FOREST DWELLERS
Many of the animal species whose remains are associated with those of *A. ramidus*, such as colobus monkeys (above) and kudu (below), eat leaves and are still usually found in forested environments today.

smaller premolars
and molars than later
australopithecines

neck bone

first rib

long arm bones
indicate that *A.
ramidus* was a
good climber

vertebra

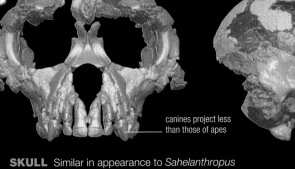

relatively
small skull

canines project less
than those of apes

small
brow

projecting
face

SKULL Similar in appearance to *Sahelanthropus* (see p.64), the *Ardipithecus* skull has a small cranial capacity, a strong but short mid-face, and relatively vertical upper jaw region. The braincase is quite low and rounded. The teeth appear to be adapted to a varied diet, with relatively small canines and molars and an enamel thickness between that of chimpanzees and *Australopithecus afarensis*.

ARDI SKULL
The bones and especially the skull were found in a very poor and highly fragmented condition. In particular, Ardi's teeth and the bones of its face were widely scattered across the excavation site.

pelvis has mix of
features useful for
both climbing and
upright walking

femur (thigh
bone)

ARMS AND HANDS The upper arm bones (humeri) from at least seven *Ardipithecus ramidus* individuals are known. The relative proportions of the arms to legs is similar to that seen in living monkeys such as macaques, which tend to walk on all fours along branches. The finger bones of *A. ramidus* are long, but the bones of the palm are short and robust and the wrist is flexible. These characteristics are consistent with clambering in the trees, which places a good deal of loading on the hands, but not with the knuckle-walking seen in modern apes.

ARDI HAND
It is rare for the small bones of the hand and foot to be preserved in fossil specimens, and these finds help establish how our ancestors made the transition from climbing to walking on two legs.

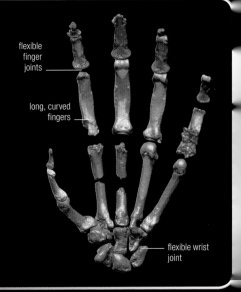

flexible
finger
joints

long, curved
fingers

flexible wrist
joint

tibia (shin bone)

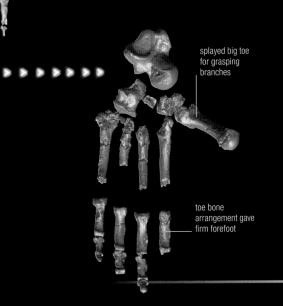

splayed big toe
for grasping
branches

toe bone
arrangement gave
firm forefoot

LOWER BODY The feet of *Ardipithecus ramidus* retained an abductable (opposable) big toe, which permitted grasping, combined with a robust and supportive mid-foot and heel that allowed for bipedal walking. The arms and legs are similar in length, which might suggest arboreal clambering, but the upper pelvis shows at least partial adaptation to upright walking. The stature and the body mass may seem large for a female, but it is thought that there was little sexual dimorphism in *Ardipithecus ramidus*.

ARDI FOOT
This well-preserved foot is particularly useful for determining how *Ardipithecus ramidus* moved, and shows an interesting mix of ape and human traits, such as an opposable big toe and a supportive mid-foot.

Ardipithecus *ramidus:*
In an area of open woodland, one individual forages in the trees for fruit and nuts while others rest and eat.

EVIDENCE FOR DIET
Scientists have been able to determine the diet of *Ardipithecus ramidus* by analyzing the thickness and molecular content of its tooth enamel. Evidence suggests that it did not eat abrasive foods but had a mixed diet dominated by woodland-based fruit, nuts, and leaves, and possibly including insects, eggs, and small mammals.

UNIQUE HANDS
The hands and wrist of *Ardipithecus ramidus* were very different from those of chimpanzees. Rather than being adapted for knuckle walking, they were better suited to supporting weight on their palms when moving around in trees. This implies that the last common ancestor of chimpanzees and humans did not move like modern apes, as was assumed by experts.

STANDING TALL
The construction of the pelvis and feet support the idea that *Ardipithecus ramidus* could comfortably stand and walk around. Nevertheless, it is thought that the walking style would have been less energy efficient than that of the australopithecines, and it is likely that the ability to run would have been limited.

ASSOCIATED ANIMALS
Although there is no direct evidence of *Ardipithecus ramidus* being preyed upon, many of the fossilized bones of mammals found in the same soil layer as "Ardi" show evidence of rodent gnawing and carnivore chewing—probably by hyenas, which certainly existed in this habitat.

TREE LIFE
Analysis of its hand, arm, feet, pelvis, and leg fossils has revealed that *Ardipithecus ramidus* would have been comfortable moving around in trees. However, they probably did not swing between branches like modern gibbons, or climb vertically like chimpanzees, but clambered about the branches supported on their feet and palms.

FAMILY GROUPS
Comparisons between the teeth of male and females have suggested that there was little difference in body size between the sexes. This may indicate that there was a low level of competition between males, and that some tasks such as food gathering and child care could have been taken on by both sexes.

Australopithecus anamensis

The remains of this hominin have a combination of characteristics found in modern apes and humans, and it is likely that it could both walk upright and climb trees.

Discovery

Most of the *Australopithecus anamensis* fossils so far described come from Kanapoi and the Lake Turkana Allia Bay site, Kenya. Since 1965, these sites have revealed the remains of several individuals, but all the fossils are fragmented. In 1994, the species was defined using KNM-KP29281, a jaw and ear canal, as the type specimen. Other equally fragmented remains have been found in Aramis, the Middle Awash, and the Omo Basin in Ethiopia. Although some anthropologists support this species, others are less confident that the range of variation among the fossils can be accounted for in a single species.

Physical features

The skull is known only from fragments, but the jaw appears narrow and forward facing, with long, parallel rows of teeth. The incisors and canine teeth protrude, and the chin recedes markedly. The ear canal is narrow, more like that of chimpanzees than modern humans. Overall the molars are modest in size with thick enamel, but tooth size varies between individuals. These characteristics suggest a diet of fruits, seeds, and leaves. A shinbone, forearm, and finger bone show evidence for weight bearing on the legs, but also on the arms. The extent of bipedalism in *A. anamensis* remains hotly debated.

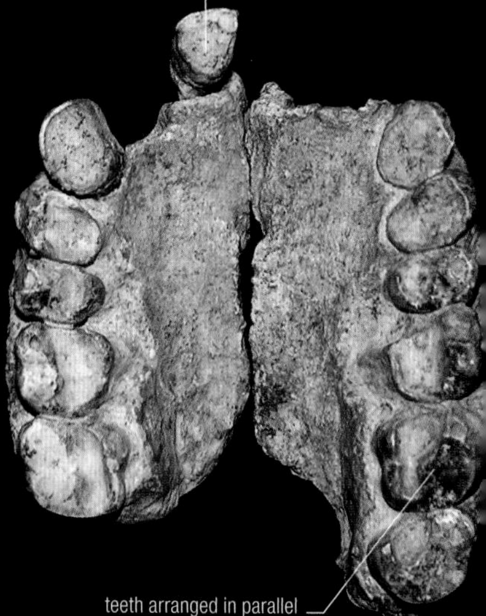

PARTIAL JAW
KNM-KP29283 is a maxilla (upper jaw) with teeth, found in 1994 in Kanapoi, Kenya, and dated to 4.15 million years ago. The bone below the front teeth slopes backward rather than projecting forward.

tooth enamel is thicker than in earlier apelike ancestors

teeth arranged in parallel rows—an apelike characteristic

AUSTRALOPITHECUS ANAMENSIS

- ▶ **NAME MEANING** "Southern ape from the Lake"
- ▶ **AGE** 4.2–3.9 mya
- ▶ **HEIGHT** Unknown
- ▶ **BRAIN SIZE** Unknown
- ▶ **LOCALITY** Lake Turkana and Kanapoi, Kenya; Aramis in Middle Awash and Omo basin, Ethiopia
- ▶ **FOSSIL RECORD** Jawbone, teeth, fragmented arm, thigh, finger bones

DESERT DIG
Excavating for hominin fossils is a slow and painstaking progress. Here, soil dug up at Allia Bay, Kenya, is carefully sieved through a fine mesh to make sure that small fragments of bone that might have been missed by excavators are recovered.

Australopithecus bahrelghazali

Australopithecus bahrelghazali is known from a sparse fossil record, but its find-spot—1,550 miles (2,500km) west of the Rift Valley—makes it an important find.

Discovery

Across most of Chad, Pliocene sediments are hidden by a thick layer of later Quaternary deposits, and therefore fossils relating to the earlier period are rare and discovered only occasionally through bore holes. However, in the region of Bahr el Ghazal lies an ancient riverbed where Pliocene sediments are exposed. In 1993, a research team led by French paleontologist Michel Brunet discovered a range of fossil vertebrates here, including a partial hominin lower jaw (KT12/H1) with seven teeth intact. Correlations with collections of faunal species from other well-dated sites suggests a date for this fossil of 3.6–3 mya.

Physical features

The type specimen (KT12/H1) is the front part of an adult lower jaw with an incisor, two canines, and four premolars. The anterior part of the jaw is quite deep and wide, with large canines compared to modern humans, and overall the specimen most closely resembles *Australopithecus afarensis* in size and proportions (see p.79). Some subtle traits have been used to distinguish it from that species, however, including the presence of premolars with three roots and relatively thin enamel. Whether this specimen is a regional variant of *Australopithecus afarensis* or a separate species, it is important because it demonstrates that early hominins had a much wider geographical distribution than previously thought.

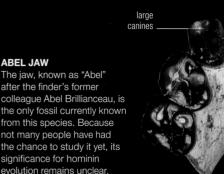

large canines — receding jawbone — incisor

ABEL JAW
The jaw, known as "Abel" after the finder's former colleague Abel Brillianceau, is the only fossil currently known from this species. Because not many people have had the chance to study it yet, its significance for hominin evolution remains unclear.

Kenyanthropus platyops

The nearly complete fossil skull that was found in 1999 could provide evidence of a flat-faced hominin living alongside *Australopithecus afarensis*.

Discovery

By the 1990s, it was clear that for much of prehistory multiple hominin species had co-existed. However, only *Australopithecus afarensis* was known from the period 4–3 mya. In 2001, a new species, *Kenyanthropus platyops*, was described from fossils found west of Lake Turkana, Kenya, and dated to the same period. These fossils included the distorted cranium KNM-WT 40000, and the upper jaw KNM-WT 38350. In 2011, stone tools dating to 3.3 mya were found nearby at Lomekwi, Kenya, suggesting that one or both of these species may have made tools long before the appearance of the genus *Homo*.

Physical features

The type specimen is KNM-WT 40000, a largely complete cranium that was distorted by geological processes. The size of the skull is within the australopithecine range, but the mid-face is flatter. The large cheekbones have a forward position, and the hard palate is wide, but the nasal opening is narrow and the earhole is small. There are few teeth preserved, but one molar is unusually small, and the enamel is quite thick. Cranial capacity is difficult to judge because of the distortion, but based on other measurements, Leakey suggested it was within the range of *Australopithecus* or *Paranthropus*.

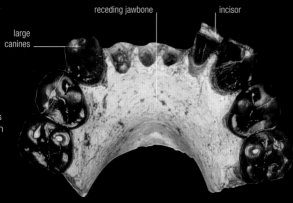

although badly deformed, skull shows a brow ridge with flat face below

poor state of skull makes teeth hard to study

DISTORTED EVIDENCE
Paleoanthropologists disagree as to whether this badly distorted skull is sufficiently well preserved to be safely interpreted as indicative of an entirely new genus living alongside the australopithecines.

Australopithecus
afarensis

It is thought that *Australopithecus afarensis* could be the ancestor of the genus *Homo,* to which modern humans—*Homo sapiens*—belong.

Australopithecus afarensis is one of the best known of the early hominins. The fragmented remains of several hundred individuals, including males, females, and juveniles, have been found in East Africa. Research on this material has revealed evidence for both terrestrial and arboreal lifestyles, and extreme sexual dimorphism.

Discovery

The Afar "triangle" in Ethiopia, where three tectonic plates meet, is well known to geologists. In the 1970s, the region became equally well known to anthropologists, because a number of extraordinarily rich fossil localitites were discovered there. In 1973, American anthropologist Donald C. Johanson made the first of his many hominin finds at Hadar—two upper thigh bones, and a knee joint that showed adaptations for upright walking. The team published their finds the following year, at the same time that Mary Leakey (see p.94) began excavating farther south at Laetoli. Johanson and his team continued their work in Ethiopia and soon discovered AL-288-1, a partial skeleton dated to 3.2 million years ago and nicknamed "Lucy." In 1977 and 1980, American anthropologist Tim White published descriptions of the Laetoli fossils found since 1974, and in 1978 Johanson, Yves Coppens, and White declared one of these specimens, the LH 4 adult mandible, as the type specimen for their new species *Australopithecus afarensis*. Over the past 20 years, many more specimens have been discovered across the region, including at Hadar, Maka, Aramis, and Dikika (Ethiopia), Laetoli (Tanzania), and west Lake Turkana (Kenya).

FOSSIL HUNTER
The team led by Donald Johanson (far left) found the first *Australopithecus afarensis* skeleton, which they christened Lucy. Later, the remains of the "first family" were discovered—13 individuals of the same species who may have died together in a flash flood.

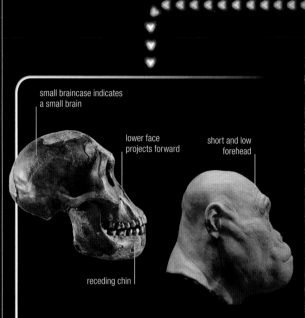

small braincase indicates a small brain

lower face projects forward

short and low forehead

receding chin

SKULL AND TEETH The large face has wide cheekbones and a long nasal region narrowly separated from the mouth. The orbits are close together and the braincase is narrow behind the eyes. The lower face and tooth row project forward, while the long upper canines are separated from the incisors by a gap into which the lower canine fitted when the mouth closed.

UPPER BODY The arms of *Australopithecus afarensis* were longer than those of modern humans. The arm bone to thigh bone ratio is similar to that of today's baboons, and relative to body size the arm bone has a large cross section, which could suggest the use of the arms for support. The fingers are longer and curved, like those of orangutans. However, the curvature of the ribs suggests a thorax that was more barrel shaped than in modern apes, and although the vertebrae have relatively small surface areas, it is likely that the upper body was held erect.

AUSTRALOPITHECUS AFARENSIS

> **NAME MEANING** "Southern ape from Afar"

> **LOCALITY** Laetoli, Tanzania; White Sands, Hadar, Maka, Belohdelie, and Fejej, Ethiopia; Allia Bay and West Turkana, Kenya

> **AGE** *3.7–3 mya*
Dated mainly through absolute dating of volcanic ash layers above and below the fossils

8 MYA · PRESENT · 7 MYA · 1 MYA · 6 MYA · 2 MYA · 5 MYA · 3 MYA · 4 MYA

> **FOSSIL RECORD** Partial adult skeleton; nearly complete baby skeleton; complete knee joint; limb fragments and other bones; several mandibles and partial crania

---

DIKIKA BABY

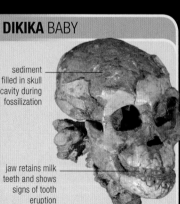

sediment filled in skull cavity during fossilization

jaw retains milk teeth and shows signs of tooth eruption

Between 2000 and 2003, a partial skeleton of an infant *A. afarensis* was recovered from deposits dated to 3.3 million years ago at Dikika, Ethiopia. The dentition and bone maturity suggest an age at death of about 3 years. The infant has long, curved fingers that raise more questions about the importance of arboreal behavior in this species.

LUCY
The specimen catalogued as AL-288-1 is nicknamed Lucy, after the Beatles song *Lucy in the Sky with Diamonds*, which was playing while the archaeologists celebrated their discovery. At the time, Lucy was the most complete fossilized hominin skeleton known. However, she was not immediately recognized as a separate species and was only later assigned to *Australopithecus afarensis*.

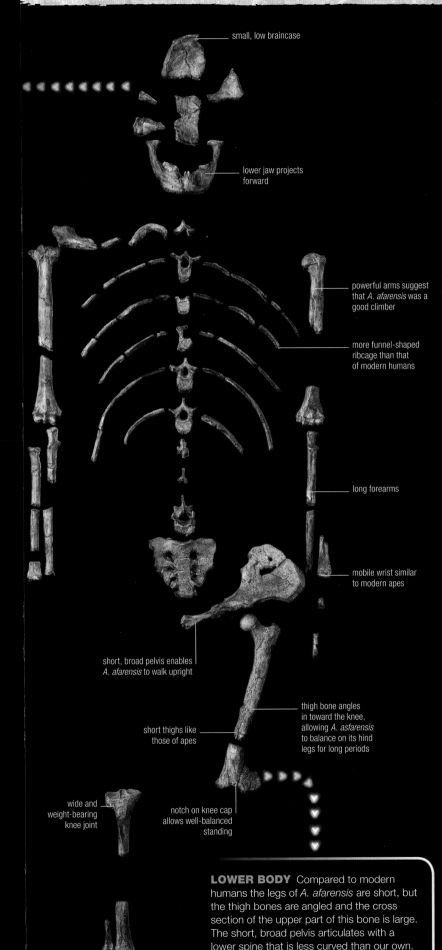

small, low braincase

lower jaw projects forward

powerful arms suggest that *A. afarensis* was a good climber

more funnel-shaped ribcage than that of modern humans

long forearms

mobile wrist similar to modern apes

short, broad pelvis enables *A. afarensis* to walk upright

thigh bone angles in toward the knee, allowing *A. asfarensis* to balance on its hind legs for long periods

short thighs like those of apes

wide and weight-bearing knee joint

notch on knee cap allows well-balanced standing

stable ankle joint

LOWER BODY Compared to modern humans the legs of *A. afarensis* are short, but the thigh bones are angled and the cross section of the upper part of this bone is large. The short, broad pelvis articulates with a lower spine that is less curved than our own. The ankle and foot bones suggest a flatter, more mobile foot, but with shorter and less opposable toes than in modern apes.

Physical features

Australopithecus afarensis is known from the remains of several hundred individuals, representing most parts of the skeleton. The braincase is small compared to the body size, but the face and jaws are large. The thorax has an erect posture, and the shape of the legs suggests the species could walk upright, while other features such as the length of the arms suggest good climbing abilities. This group of fossils includes very large and very small individuals and there is debate as to whether this variation reflects the presence of two species, or one in which males and females have very different body size.

> **HEIGHT** Males 5 ft (1.51 m); Females 3 ft 5 in (1.05 m)

> **WEIGHT** Males 93 lb (42 kg); Females 64 lb (29 kg)

> **BRAIN SIZE** 23–33 cubic in (387–550 cubic cm)

KADANUUMUU

Until 2005, Lucy was the only partial *Australopithecus afarensis* skeleton known that included both upper and lower limbs. But in January of that year, Ethiopian fossil hunter Alemayehu Asfaw, part of a team lead by Yohannes Haile-Selassie, discovered another example at Korsi Dora vertebrate locality 1, in Ethiopia's Afar region. This skeleton—dated to 3.58 million years ago and known as Kadanuumuu, or "big man"—provides new evidence on stature, limb proportions, and locomotion in *A. afarensis*. Kadanuumuu was probably a male, and is much larger bodied than Lucy. This larger stature is supported by footprints found at Laetoli in 2015, which were made by individuals of a similar height. Although Kadanuumuu's lower limbs are rather short in comparison to some later hominins, in many respects his skeleton is highly adapted to effective upright walking. Even the shoulder blade, or scapula—which is a rare find in the fossil record—shows little evidence for hanging in the trees or vertical climbing. The shape of the ribs suggests that the thorax was less funnel shaped than previously assumed, and the legs are angled under the body to improve balance.

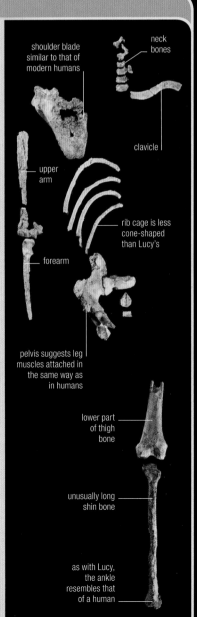

shoulder blade similar to that of modern humans

neck bones

clavicle

upper arm

forearm

rib cage is less cone-shaped than Lucy's

pelvis suggests leg muscles attached in the same way as in humans

lower part of thigh bone

unusually long shin bone

as with Lucy, the ankle resembles that of a human

SKELETON OF A BIG MAN
The skeleton of Kadanuumuu is missing its skull and teeth. As these elements are some of the most crucial for establishing to which species a specimen should be assigned, some people disagree with its identification as *Australopithecus afarensis*. However, at the moment most researchers agree that Kadanuumuu is a close relative of Lucy and the Dikika baby.

Archaeology

Hominin fossils are not the only source of information about the ancient past. The archaeological record also includes many other kinds of physical evidence, such as stone tools, processed materials, animal and plant remains, and changed landscapes. Paleoanthropological fieldwork is a collaborative process in which experts in all these disciplines come together to interpret a site and the finds within it. Two examples of this are the study of footprints and animal remains.

Preserved footprints

In 1974, Mary Leakey began a new phase of survey at Laetoli, where she had started work in 1935. New fossil finds included the type specimen for *Australopithecus afarensis* and also a series of remarkable footprints. Covered by a layer of volcanic ash that fell 3.7 million years ago, many species—including hominins—had left footprints in wet ashfall that subsequently hardened. The hominin tracks appear to have been made by two individuals, one larger than the other, with a probable third individual walking behind. These footprints provided valuable information about how early hominins walked. The slightly divergent big toe reflects a more mobile foot than seen in modern humans, but the deeper heel and toe impressions point to a heel-strike and toe push-off similar to modern walking. In 2015, more footprints were found; most were made by a single *A. afarensis* individual who, at an estimated 5ft 5in (1.65m) tall, is around the same height as Kadanuumuu (see p.79), suggesting many members of this species were taller than had been thought.

FAMILIAR FOOTPRINT
The Laetoli footprints preserve evidence of the way the *afarensis* individuals' bodyweight was distributed through their feet while walking. Taken together with the evidence of the foot bones, the prints demonstrate how similar their gait was to our own, rather than to that of a chimpanzee walking upright.

upper part of joint angles towards the hip, indicating upright walking

top of shin bone is widened to bear more weight, an essential feature for bipedalism

COMPLETE KNEE JOINT
It is very rare that both parts of the knee joint are preserved in a fossil specimen, and this find has given us valuable information on how *A. afarensis* walked.

STEPS BACK IN TIME
Australopithecus afarensis is not the only species whose footprints are preserved by the volcanic ash at Laetoli. Visible in the top right of the image (opposite) are prints made by an extinct type of three-toed horse called *Hipparion*.

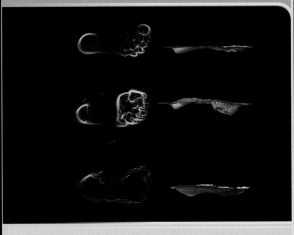
Evidence of early tool use

Until recently, the earliest evidence for stone tools was that of the Oldowan tradition (see p.102), dating from around 2.6 million years ago in East Africa. However, new finds suggest that stone tool technology is much older. Stone tools from a site called Lomekwi 3 at Lake Turkana, Kenya, date to 3.3 million years ago. They are so distinct from later Oldowan tools that the excavators have assigned them to a separate tradition, the Lomekwian. Other, less direct evidence from locality DIK-5, in the Lower Awash Valley, Ethiopia, also suggests early tool use around 3.4 million years ago. Cutting-edge imaging technology has revealed that marks on two fossil ungulate bones discovered in 2009 were made before fossilization, possibly by stone tools being used to remove meat from the bones and smash them open for marrow. These new early dates precede the appearance of the genus *Homo*, suggesting that australopithecines and other early genera may also have made stone tools.

marks A1 and A2 may have been made by a sharp-edged stone tool

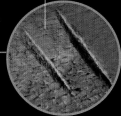

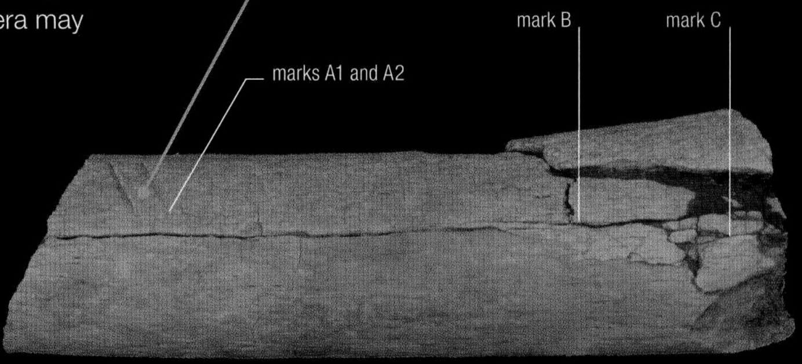

mark B

mark C

marks A1 and A2

BUTCHERED BONE?
It has been suggested that marks A1 and A2 may have been made by a cutting tool, and marks B and C by a hammerstone. If true, this would put the date for the earliest known tool use back by about 1 million years.

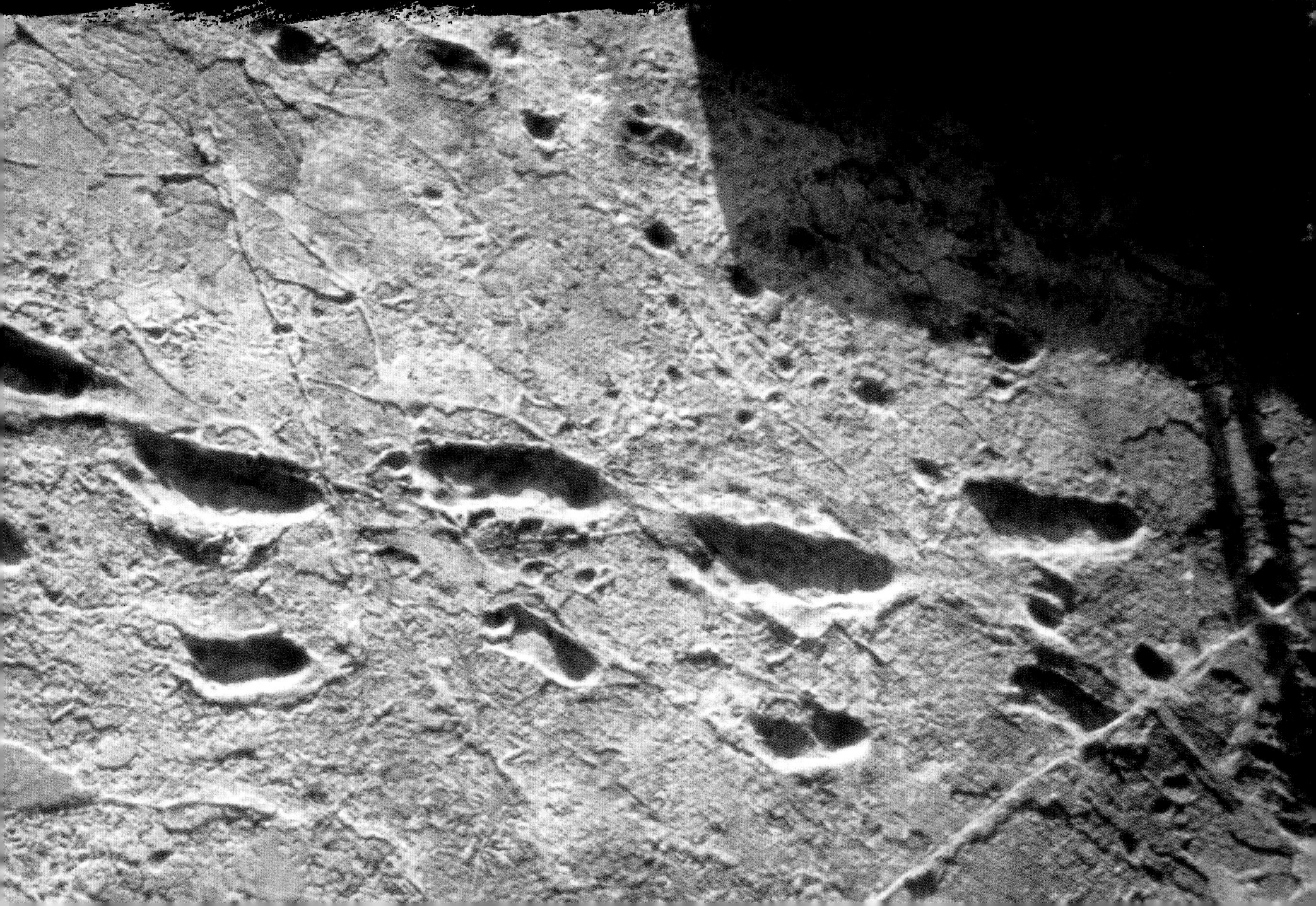

VOLCANIC ERUPTION
The Sadiman volcano is located some 12 miles (20km) east of the site where the Laetoli footprint tracks were laid down. It was very active 3.6 million years ago, and frequent eruptions had formed a conical-shaped profile. Today, the volcano is long extinct and its slopes are greatly eroded.

SIZE DIFFERENCES
The Laetoli footprints are those of different-sized individuals walking together. One individual was up to 5ft 5in (1.65m) tall, while the others were less than 4ft 9in (1.45m). This striking size difference may mean there was significant sexual dimorphism—size differences between the sexes—among *Australopithecus afarensis*. However, others believe that the footprints were made by more than one species.

Australopithecus *afarensis:*
A small group walks through a desolate landscape leaving tracks of footprints in recently fallen volcanic ash.

MYSTERIOUS FIGURE
Although it is clear that the tracks were made by two individuals walking side-by-side, many of the larger footprints forming the right-hand track appear blurred. Some scientists suggest that this distortion could have been created by a third individual stepping in the tracks of the larger individual ahead.

ASSOCIATED ANIMALS
The habitat of this part of East Africa was similar to that existing today, and so it is not surprising that the animal life was also similar. More than 20 species of savanna-based animals left their tracks alongside *Australopithecus afarensis*, including giraffes, antelopes, rhinoceroses, buffalo, and elephants. The gigantic *Deinotherium* is shown here.

FOOTPRINT FORMATION
The Laetoli footprint tracks were formed by a specific sequence of events. An initial eruption left a fine layer of powdery ash through which the hominins left their mark. A soft rain shower, followed by sunshine, had the effect of solidifying this layer, before subsequent eruptions covered the tracks until they were eventually unearthed.

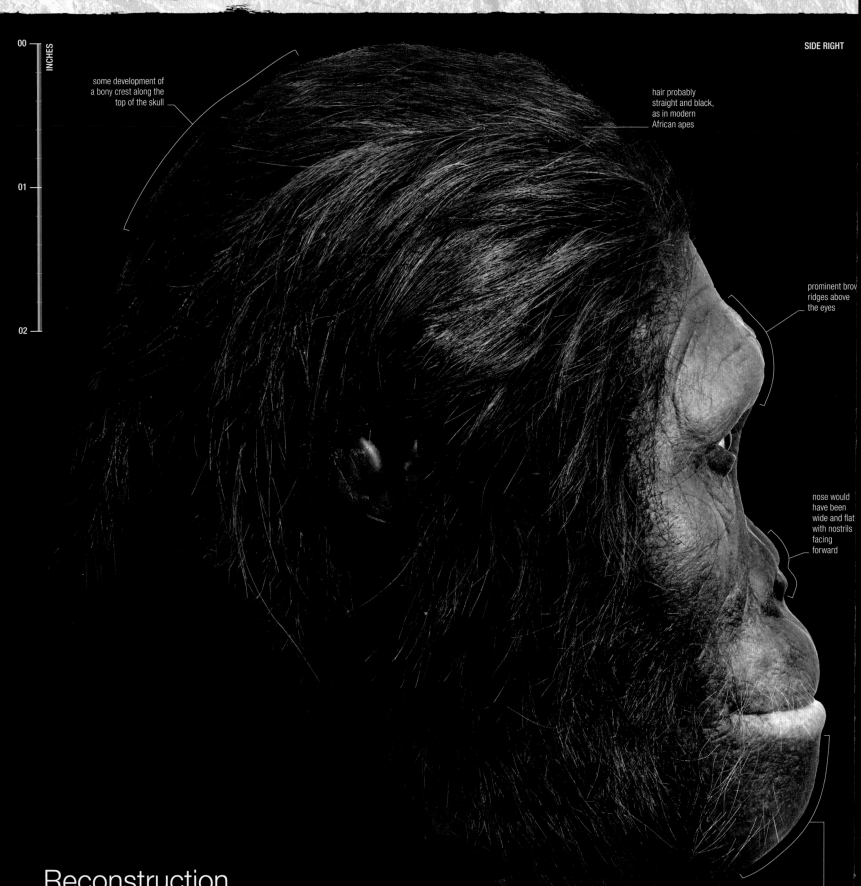

00

INCHES

01

02

some development of
a bony crest along the
top of the skull

hair probably
straight and black,
as in modern
African apes

prominent brov
ridges above
the eyes

nose would
have been
wide and flat
with nostrils
facing
forward

deep lower
jaw lacks a
distinct chin

Reconstruction

This reconstruction of an adult male was based on
pieces of skull and jaw found among a group of fossils
known as the "first family," which was made up of 17
individuals of various ages. This hominin had a distinctly primitive appearance
when compared to later *Homo* species, and it retained many characteristics
of our Miocene ape ancestors. Its braincase was very small and it had a wide,
dish-shaped upper face with a flat nose and forward-facing nostrils. As in
modern apes, its lower face was prognathic (with jaws that projected outward).

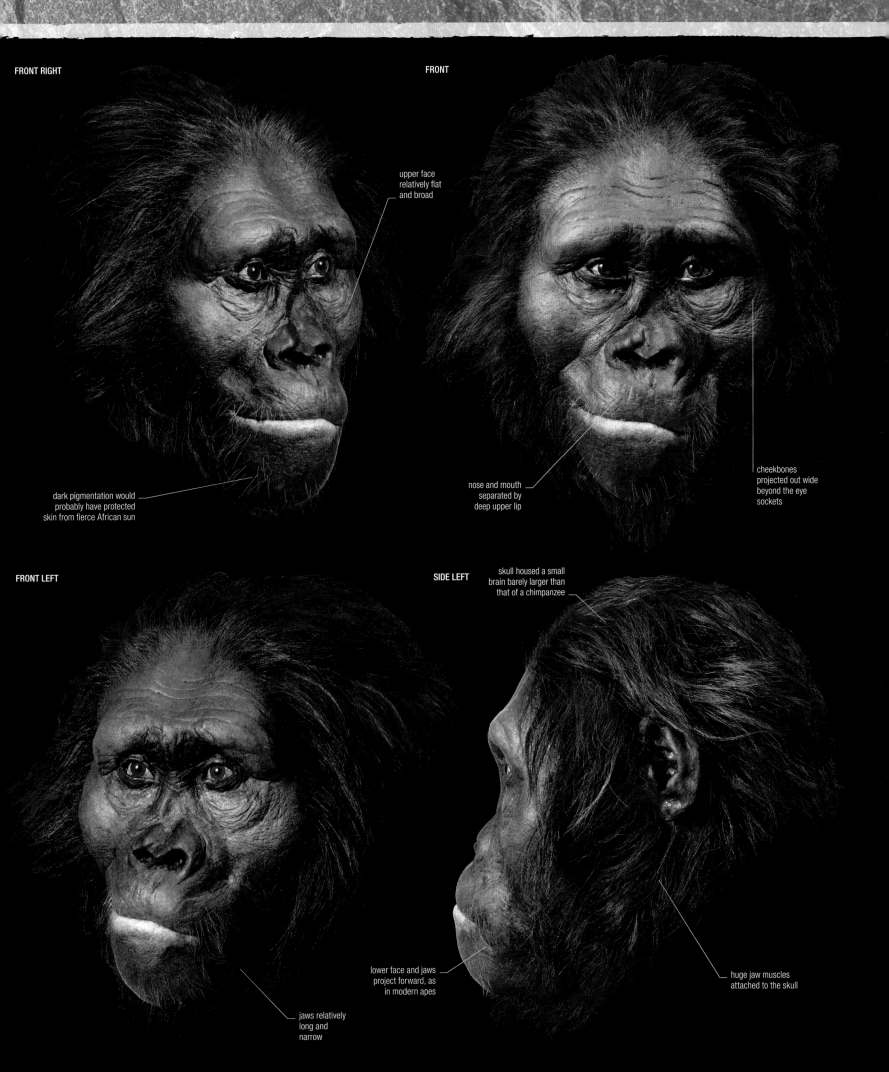

FRONT RIGHT

FRONT

upper face
relatively flat
and broad

dark pigmentation would
probably have protected
skin from fierce African sun

nose and mouth
separated by
deep upper lip

cheekbones
projected out wide
beyond the eye
sockets

FRONT LEFT

SIDE LEFT

skull housed a small
brain barely larger than
that of a chimpanzee

lower face and jaws
project forward, as
in modern apes

huge jaw muscles
attached to the skull

jaws relatively
long and
narrow

Australopithecus
africanus

The first early hominin species to be identified,
Australopithecus africanus firmly placed the
location of ancient hominin evolution in Africa.

Australopithecus africanus was the species that proved fossil hominins were present in Africa. Its discovery showed that, despite having small brains, early hominins could walk upright.

Discovery

Although Charles Darwin had argued that human origins probably lay in Africa, until the late 1920s fossil hominins had only been found in Europe and Asia. All this changed in 1924, when a box of fossils and rocks was sent to Raymond Dart, the Professor of Anatomy at South Africa's Witwatersrand University. The collection had come from the Buxton limeworks at Taung, on the edge of the Kalahari Desert. Dart was a skilled neuroanatomist and recognized that one of the specimens was an unusual primate "endocast," or natural cast of a braincase. It matched a partial juvenile skull, and with its vertical forehead, smooth brow, and slender bones, Dart quickly concluded that it represented an "extinct race of apes intermediate between living anthropoids and man." In 1925, Dart named this species *Australopithecus africanus*. Since then, many other fossils have been found in cave deposits at Sterkfontein and Makapansgat, including nearly intact skulls and skeletons.

RAYMOND **DART**

Raymond Arthur Dart was born in Queensland, Australia, in 1893. After studying biology at the University of Queensland and medicine at the University of Sydney, he traveled to England to serve in the Medical Corps during World War I. In 1923, Dart took the position of Professor of Anatomy at the University of Witswatersrand in Johannesburg. A year later, he identified the oldest infant hominin skull ever found—the "Taung child." Stung by criticism of his work and the rejection of his claims about *Australopithecus africanus*, Dart's paleoanthropological activity waned. He died at the age of 95, having lived to see his views on the Taung child vindicated.

AUSTRALOPITHECUS AFRICANUS

> **NAME MEANING** "African southern ape"

> **LOCALITY** Limestone caves in Sterkfontein, Makapansgat, Taung, and Gladysvale, South Africa

> 8 MYA   **PRESENT**   1 MYA
> 7 MYA
> 6 MYA   2 MYA
> 5 MYA   3 MYA
> 4 MYA

AGE *3.3–2.1 mya*
Dated through relative dating based on matching fossils found in caves with fossils from absolutely dated sites in East Africa

> **FOSSIL RECORD** Several partial skulls, a number of jawbones, various skeleton fragments

STERKFONTEIN CAVES
These caves, northwest of Johannesburg, are among the most important fossil sites known. They have produced finds of australopithecines, paranthropines, and early *Homo*. The caves are now a UNESCO World Heritage Site.

Physical features

While many of the specimens are fragmented or distorted, nearly all parts of the skeleton are represented by the *Australopithecus africanus* fossil record. The evidence suggests a hominin of modest body size, capable of upright walking, with a pattern of growth and maturation more similar to modern apes than humans. Some adult skulls appear to have been much larger (Sts 19) than others (StW 505), which may indicate differences between the sexes and perhaps a haremlike social organization similar to that seen in modern gorillas. Alternatively, this variability in skull size may represent two different groups.

> **HEIGHT** Female: 3ft 7in (1.10m); Male: 4ft 5in (1.35m)
>
> **WEIGHT** 55–110lb (25–50kg)
>
> **BRAIN SIZE** 26–38 cubic in (428–625 cubic cm)

UPPER BODY The relatively long arms, the mobile shoulder, and long, large hand bones indicate a load-bearing upper body. *Australopithecus africanus* was probably an adept climber, and would have held its trunk upright when feeding.

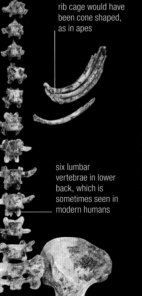

rib cage would have been cone shaped, as in apes

LOWER BODY Like *Australopithecus afarensis*, the pelvis, femur (upper leg), and foot bones of *Australopithecus africanus* indicate that it could comfortably walk bipedally. However, the toes are long, and the foot is more mobile than that of modern humans, with a flexible arch and a more divergent big toe. The lower vertebrae of the spine also have smaller suface areas than in modern humans, perhaps suggesting a different range of motion or weight-bearing capacity.

six lumbar vertebrae in lower back, which is sometimes seen in modern humans

pelvis adapted for bipedalism, but less rounded than in modern humans

PELVIS AND SPINE
The pelvis and spine (Sts 14) of *A. africanus* found at Sterkfontein by Robert Broom in 1947 indicated that the species was bipedal, and challenged the idea that our ancestors' brains grew significantly before they walked on two legs.

SKULL The skull of *Australopithecus africanus* is rounded and lightly built. The brain volume, at around 27 cubic in (450 cubic cm), is similar to the mean values of modern great apes, but the cranial vault is domed and the muscle markings are faint. The neck-muscle attachments are low on the back of the skull, which sits on a vertical spine. On the face, the brow ridge is modest and the cheekbones are thin, but the upper jaw is broad and projecting. The front teeth are correspondingly large, and the canines show slight sexual dimorphism.

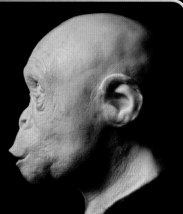

TAUNG CHILD
Australopithecus africanus closely resembled *A. afarensis*, but had a larger jaw and strong cheekbones to support the chewing muscles. The cheek teeth were also bigger, but resembled those of humans more than earlier hominins.

gracile cheekbones

domed braincase

broad upper jaw

"MRS. PLES"
The most complete skull (Sts 5), found by Robert Broom at Sterkfontein in 1947, was initially classified as a *Plesianthropus transvaalensis* female (hence its nickname). It is now thought to have belonged to a young male.

brain size similar to that of modern apes

TAUNG CHILD

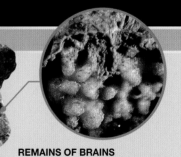

skull has a mixture of human and apelike features

first molar teeth were found to be erupting in the jaw

REMAINS OF BRAINS
During fossilization, the child's skull filled with sediments saturated with calcium-rich water. The sediments hardened into stone, preserving a perfect replica of the inside of the skull.

The Taung child was the first *A. africanus* fossil to be found, and it is the type specimen for the species. Assessing the age of death in extinct species is difficult since the rate of growth and maturation is unknown, but comparisons with living primates provide clues. The Taung child retains all its milk teeth, and the first permanent molars are just erupting. This pattern appears at about 6 years in humans and about 2–3 years in great apes. Microscopic analysis of the tooth enamel and bone formation rates suggest a more apelike pattern, so the Taung child is likely to have been 2–3 years old when it died.

Taung child has no
significant brow ridges
while even young apes
show some bony ridges
over the eyes

Reconstruction

Although *Australopithecus africanus* shared many similarities with later
Homo species, it may be more closely related to the paranthropines.
This reconstruction is based on the skull of the Taung child, who is thought
to have died aged about 2–3 years old between 2.5 and 2 million years ago. This
skull already shows many of the species' key traits, but distinctively "apelike"
features often develop during growth in modern apes, and young apes look
more like humans than adults do. So the Taung child must be compared with adult
specimens in order to reach firm conclusions about the species' characteristics

jaw does not project very
far, although this may be
because the Taung child
was so young; this feature
develops quite late in apes

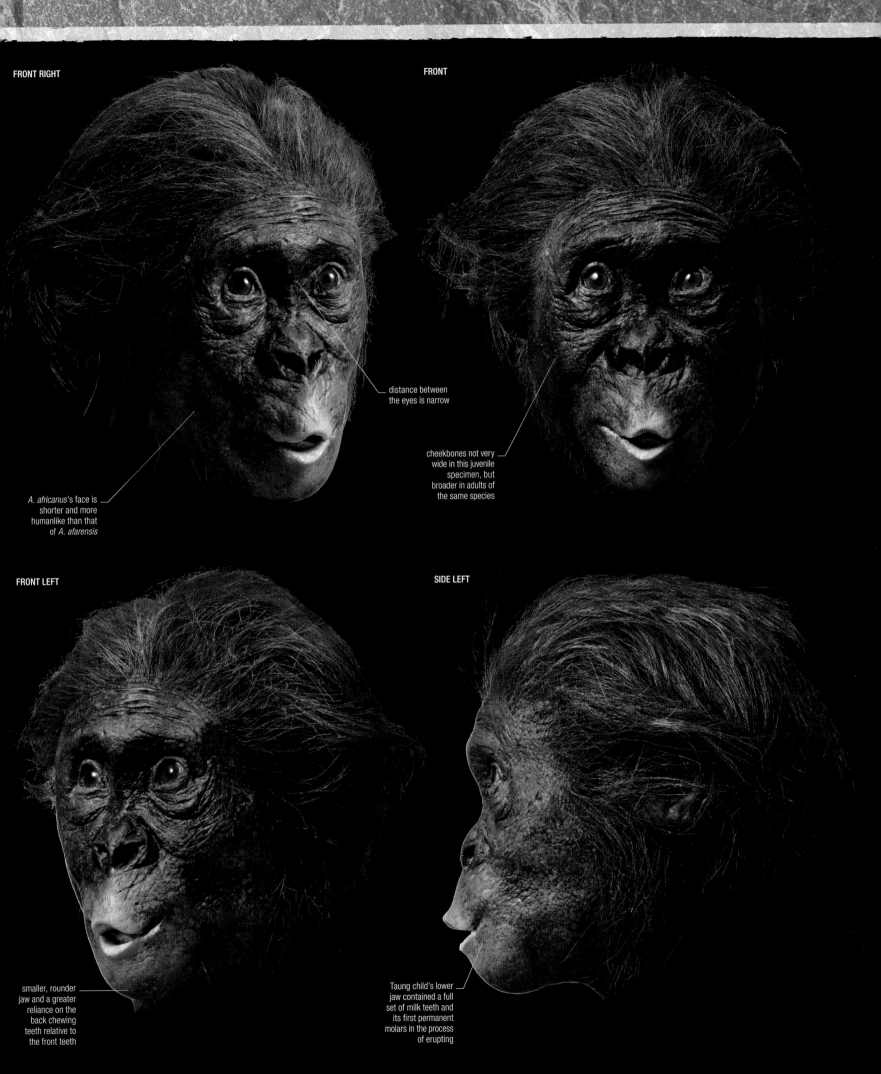

FRONT RIGHT

FRONT

distance between
the eyes is narrow

cheekbones not very
wide in this juvenile
specimen, but
broader in adults of
the same species

A. africanus's face is
shorter and more
humanlike than that
of *A. afarensis*

FRONT LEFT

SIDE LEFT

smaller, rounder
jaw and a greater
reliance on the
back chewing
teeth relative to
the front teeth

Taung child's lower
jaw contained a full
set of milk teeth and
its first permanent
molars in the process
of erupting

Australopithecus garhi

Found in association with grassland and scrubland animal species, this hominin is a candidate ancestor for early *Homo*.

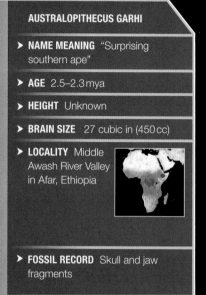

AUSTRALOPITHECUS GARHI
NAME MEANING "Surprising southern ape"
AGE 2.5–2.3 mya
HEIGHT Unknown
BRAIN SIZE 27 cubic in (450 cc)
LOCALITY Middle Awash River Valley in Afar, Ethiopia
FOSSIL RECORD Skull and jaw fragments

Discovery

In 1990, a team led by anthropologists Berhane Asfaw (Ethiopia) and Tim White (USA) recovered hominin fossils from the Bouri Formation in the Middle Awash River Valley, Ethiopia. A mandible that retained tooth roots suggested, surprisingly, that although quite large and dating to around 2.5 MYA, this was not *Paranthropus aethiopicus*. In 1996–98, more remains were found at Bouri. These included parts of the arms and thigh bones of two to three individuals, and a partial skull (BOU-VP-12/130) that showed a distinctive suite of characteristics and which became the type specimen for a new species *Australopithecus garhi*—the "surprise."

Physical features

Australopithecus garhi was a hominin of modest body size. There is some evidence that it had longer legs than any of the other australopithecines, suggesting a more *Homo*-like walking mode. At the same time, it appears that the arms were longer, and more robust, than in later hominins. Running along the midline of the braincase was a crest of bone, suggesting that *A. garhi* had strong chewing muscles. Long-bone size is variable, which could suggest sexual dimorphism (difference between the sexes). Some anthropologists have remarked that the cranial characteristics are similar to those of *A. africanus* (see p.88).

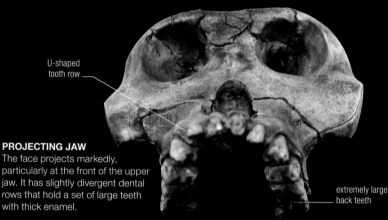

U-shaped tooth row

extremely large back teeth

PROJECTING JAW
The face projects markedly, particularly at the front of the upper jaw. It has slightly divergent dental rows that hold a set of large teeth with thick enamel.

Paranthropus aethiopicus

The first fossil hominin to be found in Ethiopia, but now best known from Kenyan sites, this species shows a remarkably robust skull.

PARANTHROPUS AETHIOPICUS
NAME MEANING "Ape that lived alongside humans, from Ethiopia"
AGE 2.7–2.3 mya
HEIGHT Unknown
BRAIN SIZE 25 cubic in (410 cc)
LOCALITY Lake Turkana, Kenya; Omo, Ethiopia
FOSSIL RECORD Two skulls, sections of jaw, and various teeth fragments

Discovery

The first hominin species to be found in Ethiopia was uncovered in 1967 and named *Paraustralopithecus aethiopicus*. This partial jawbone, without teeth, did not make a huge impact on anthropology at the time. Nearly 20 years later, at West Turkana in Kenya, a team led by Kenyan anthropologist Richard Leakey uncovered the well-preserved and distinctively wide-faced "Black Skull." By 1987, this was grouped with the Ethiopian find as *Paranthropus aethiopicus*. Several other collections of fragmented jaws and teeth have also been ascribed to this species, including material from the Shungura Formation in Kenya.

Physical features

Paranthropus aethiopicus is known almost exclusively from teeth and skull fragments. Like the other paranthropines, it has small canines and large molars that wear flat, which may suggest that at times this species needed to eat foods requiring tough chewing. Unlike other paranthropines, the back tooth rows are long and parallel, and the front teeth are wide. The face tends to be large, projecting, and slightly dished (concave) in profile. Behind the modest brow ridge, the skull narrows and the cranial capacity is small, at around 25 cubic in (410 cubic cm).

distinct crest

dish-shaped mid-face

lower face projects outward

cheekbones positioned outward

THE BLACK SKULL
The best-preserved *Paranthropus aethiopicus* specimen is KNM-WT-17000, known as the "Black Skull." It has a dramatically wide face set in front of a small braincase, and huge sagittal crest for the attachment of chewing muscles.

Paranthropus robustus

A very robust hominin, the South African *Paranthropus robustus* had massive back teeth and was well adapted to a diet of tough or fibrous foods.

Discovery

Paranthropus robustus was discovered at Kromdraai, South Africa, by British-born paleontologist Robert Broom in 1938. Broom had purchased some fossil fragments from locals and insisted on being shown the findspot. There he found a very robust braincase, face and jaw, ankle bone, and an elbow joint. Convinced that these represented a hominin outside of the direct human lineage, he published the material as *Paranthropus robustus*. Broom went on to excavate more robust fossils from Swartkrans between 1948–49. More recently, the very well-preserved skull of a female (DH7) was found at Drimolen in 1992.

PARANTHROPUS ROBUSTUS	
▶ **NAME MEANING**	"Robust ape that lived alongside humans"
▶ **AGE**	2–1.2 mya
▶ **HEIGHT**	3 ft 7 in–4 ft 3 in (1.1–1.3 m)
▶ **BRAIN SIZE**	32 cubic in (530 cc)
▶ **LOCALITY**	Kromdraai, Swartkrans, Gondolin, Drimolen, and Cooper's caves in South Africa
▶ **FOSSIL RECORD**	Various skulls; jaw, teeth, and skeleton fragments

KROMDRAAI EXCAVATION
Kromdraai is a breccia-filled cave site where Broom found the first *P. robustus* fossils. Together with other sites, such as Sterkfontein and Swartkrans, it is part of a UNESCO World Heritage site.

Physical features

Paranthropus robustus is known from a collection of largely fragmented fossils. Although probably modest in body size, with males rather larger and more robust than females, they had distinctively wide and slightly dished faces, with broad cheekbones and a strong sagittal crest along the braincase—all features associated with anchoring massive chewing muscles to the skull and jaw.

OVERSIZED TEETH
The molars and premolars of *P. robustus* were huge and thickly enameled, often wearing flat. These adaptations would have allowed these hominins to eat hard, fibrous foods.

Australopithecus sediba

Australopithecus sediba may show part of the hominin transition from arboreal locomotion to a more modern pattern of bipedal walking.

Discovery

In August 2008, 9 year-old Matthew Berger was on site at Malapa, South Africa, with his father, paleoanthropologist Lee Berger (see pp.158–59), when he found the collarbone of an immature australopithecine. During excavation, the research team found several partial skeletons from individuals that had apparently fallen to their death in a pool at the bottom of the Malapa Cave. Rapid burial in cave sediments had ensured that the fossils were very well preserved. The remains of plants trapped in dental plaque were found on some of the fossils' teeth, and even portions of what may be skin had been preserved.

AUSTRALOPITHECUS SEDIBA	
▶ **NAME MEANING**	"Southern ape from the spring or well"
▶ **AGE**	1.977 mya
▶ **HEIGHT**	4 ft 2 in (1.27 m)
▶ **BRAIN SIZE**	26–27 cu in (420–450 cc)
▶ **LOCALITY**	Malapa fossil site (near Johannesburg), South Africa
▶ **FOSSIL RECORD**	Two partial skeletons and numerous fragments

Physical features

Australopithecus sediba shows an intriguing combination of characteristics. It is similar to the other australopithecines in its small body size, long upper limbs, curved finger bones, and small brain. However, it also shares some features with *Homo* species. The face is narrow, the brow ridge is small, the lower jaw is quite pronounced, and the chewing teeth are relatively small, while males and females do not differ much in size. Increased bony stabilization of the hip joint and a strong thigh bone suggest human-like locomotion. The microscopic plant remains called phytoliths found on *A. sediba*'s teeth indicate that it ate a surprisingly wide diet including foods from both forest and open environments.

rounded skull

MOSAIC OF FEATURES
This is the skull of a male about 12–13 years old, with a mosaic of primitive and more recent characteristics that shows a link between *Homo* and australopithecines.

teeth similar in size to those of *Homo* species

Paranthropus boisei

Paranthropus boisei, often nicknamed "nutcracker man," was a large parathropine with distinctive massive jaws and cheek teeth, and very strong muscles and bones associated with chewing. The species was sexually dimorphic, with males much larger than females. Found in East Africa, it seems that *Paranthropus boisei* was successful and persisted for about 1 million years.

Discovery

After many years of fieldwork at Olduvai Gorge, Tanzania, Louis and Mary Leakey were rewarded with one of the key finds of their time: an extremely robust skull that defined a new species of hominin. When unearthed, the find revealed the near-complete cranium of an adult, including large teeth still articulated in the jaw. Catalogued as Olduvai Hominin 5 (OH 5), the new specimen was named *Zinjanthropus boisei* (East African man) by Louis Leakey, in recognition of both the findspot and the support of his sponsor, Charles Boise. Not only was this specimen the first of its kind to be found, but at the time it was also the earliest known hominin from East Africa. Pioneering geochemical dating techniques were used on volcanic materials from the site to determine the geological age of the findspot; this had never been attempted at a hominin-bearing site before. When this method, called potassium/argon (K/Ar) dating, was first used at Olduvai most anthropologists thought that human evolution stretched back only about half a million years. The results

LOUIS AND MARY **LEAKEY**

Louis Leakey was a Kenyan paleoanthropologist who became a curator of the Coryndon Museum (later the Kenya National Museum) in Nairobi. In 1936, he married English archaeologist Mary Nichol. For nearly 30 years the pair undertook joint excavations and surveys at Olduvai, describing the Oldowan tool industry (see p.102) and discovering the "Zinj" skull. From the 1950s, Mary concentrated on the archaeology of Olduvai, while Louis became involved in other projects, such as primatological fieldwork in Africa and Asia.

for the "Zinj" site staggered the scientific community, demonstrating that the skull was around 1.75 million years old. Now placed in the *Paranthropus* genus alongside other robust hominin species, the specimen still dubbed "Zinj" retains a special place in the story of paleoanthropology.

PARANTHROPUS BOISEI

> **NAME MEANING** Named after sponsor, Charles Boise

> **LOCALITY** Olduvai and Peninj, Tanzania; Omo Shungura Formation and Konso, Ethiopia; Koobi Fora, Chesowanja, and West Turkana, Kenya

>

8 MYA	PRESENT
7 MYA	1 MYA
6 MYA	2 MYA
5 MYA	3 MYA
	4 MYA

AGE *2.3–1.4 mya*
Dated mainly from absolutely dated layers of volcanic ash above and below the sediments bearing the fossils

> **FOSSIL RECORD** Several well-preserved skulls and crania; many jaws and isolated teeth. No body or limb fossils confirmed for this species

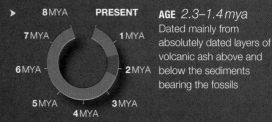

SITE OF ORIGINAL FIND
A plaque marks the spot in Olduvai Gorge, Tanzania, where Mary Leakey discovered the "Zinj" skull. She found a jawbone during a routine site inspection. The cranium was subsequently unearthed and reconstructed from its fragments.

Physical features

Fossil skulls and teeth show *Paranthropus boisei* to have been a robust species, characterized by a broad, long face, extremely large molar teeth, and strongly developed bones of the braincase. *P. boisei* skulls are large relative to those of other early hominins, with a mean brain capacity of 31 cubic in (508 cubic cm). Many anthropologists have noted that these characteristics are very similar to those of *Paranthropus robustus* (see p.93), although they are typically more exaggerated in *P. boisei*.

> **HEIGHT** Males: 4 ft 6 in (1.37 m); Females: 4 ft (1.24 m)

> **WEIGHT** Males: 108 lb (49 kg); Females: 75 lb (34 kg)

> **BRAIN SIZE** 29–33 cubic in (475–545 cubic cm)

NUTCRACKER MAN
P. boisei was nicknamed "Nutcracker man" because of its powerful jaws and large teeth. The skull resembles that of *P. robustus*, but it is bigger and more specialized for heavy chewing.

CRANIUM The cranium of *P. boisei* is large, with broad cheekbones and a robust maxilla (upper jaw). The eye orbits are rounded and very widely spaced, surrounded by a pronounced brow region. The face is relatively flat and slightly concave, and the whole skull is short front to back, with strong attachment areas for the neck and chewing muscles both on the base and running along the top in a sagittal crest.

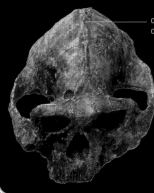

crest runs down the center of the skull

cheekbones form wide arches

TOP VIEW
Seen from above, *P. boisei*'s huge, flared cheekbones and narrow braincase (especially just behind the eye sockets) are very apparent.

PARANTHROPINE DIET

The robust nature of the skulls of paranthropine species, including very large jaws and teeth, has led experts to believe that paranthropines had a very limited diet of hard, tough foods like grasses, sedges, nuts, and seeds that needed heavy chewing. However, a study of microscopic marks on paranthropine teeth (including those shown above) and chemical analysis of fossils suggests that paranthropines had a much broader diet, including fruit and, possibly, meat. Their robust skulls would have enabled them to "fall back" on the tougher foods during lean seasons.

thick tooth enamel

strongly built lower jaw

lower face projects less than in other early hominins

BODY AND LIMBS
Apart from skulls, very few fossils of other skeletal regions are confidently ascribed to *P. boisei*. A small number of skulls have associated noncranial fragments that suggest the body size of this species varied substantially, and that the femur was adapted for bipedalism. Until more material is identified, the evidence is inconclusive.

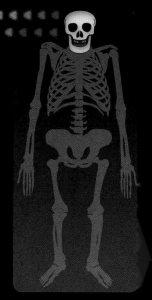

JAWS AND TEETH The upper and lower jaws (maxilla and mandible) are big and sturdy, to accommodate the huge molars and premolars that are among the largest found in any hominin species. The maxilla is wide and deep, while the mandible lacks a distinct chin, as in other early hominins. In many *P. boisei* specimens, the cheek teeth have been worn flat, which may indicate a diet that included abrasive foods. In contrast to the rear teeth, the incisors and canines are small. The canines have flat wearing tips similar to the other teeth and do not vary significantly in size between specimens.

small front teeth

molars are among the largest of any hominin species

EVIDENCE OF DIET
Although the jaws and teeth could easily crush hard, brittle foods such as nuts and tubers, the microscopic wear patterns on the teeth resemble those of fruit-eating primates. *P. boisei* probably ate nuts and tubers when its preferred food was hard to find.

NTHROPUS BOISEI

length of 10½ in
this male's head is
antly larger than the
s, which is 8½ in (21 cm) long.
oust head is reconstructed from
5 skull found at Olduvai Gorge.

FEMALE PARANTHROPUS BOISEI
This reconstruction is based on the
KNM-WT 17400 cranium that was
found in West Turkana, Kenya.
It was from a female *P. boisei* who
lived 1.7 million years ago.

Sexual dimorphism

In addition to differences in the sex organs, adult males and females of the same species often also differ in other ways, such as body size, shape, and coat color. Some species, such as gibbons and modern humans, show little sexual dimorphism while others, such as modern gorillas, are highly dimorphic, with males weighing up to twice as much as females. Similarly, specimens of *Paranthropus boisei* that are much larger than others, and often have sagittal crests (see p.95), are probably those of males, while the smaller, crestless crania are probably those

of females, who were much less heavy and muscular and also had smaller teeth than the males. High levels of sexual dimorphism usually evolve in species whose males compete with each other for access to females, and therefore benefit from being large and intimidating. This suggests this form of male-female relationship characterized *P. boisei*. In contrast, like modern humans, *Homo ergaster* and later *Homo* species are much less sexually dimorphic, perhaps suggesting that males and females had less competitive mating systems or even formed pair-bonds.

Homo *habilis*

This species was named *Homo habilis*, meaning "handy man," because some of its fossils are associated with early stone tools.

Homo habilis is the earliest member of the genus *Homo* to appear in the fossil record. This species is characterized by a slightly larger brain, smaller molars and premolars, and more human-like feet than those of earlier hominins.

Discovery

In the early 1960s, fossil fragments were excavated at Olduvai Gorge, Tanzania, from deposits slightly older than those in which "Zinj" (*P. boisei*, see p.94) had been found in 1959. They included a partial cranium, mandible, hand bones, and a near-complete left foot. The first three were grouped as specimen OH7, the foot as OH8. In 1961, Louis Leakey concluded that these finds were a different species from "Zinj," one more closely related to modern humans and capable of making the stone tools found at Olduvai. In 1964, after more fossil discoveries, Leakey, paleoanthropologist Phillip Tobias, and paleontologist John Napier described the collection as *Homo habilis*.

shorter trunk than
a modern elephant

DEINOTHERIUM
Remains of this elephant-like mammal, which was around 13 ft (4 m) high at the shoulder, have been found at all the major sites in East Africa where hominids have been found, including at Olduvai.

HOMO HABILIS

> **NAME MEANING** "Handy man"

> **LOCALITY** Olduvai Gorge, Tanzania; Koobi Fora, Kenya; Omo and Hadar, Ethiopia; Sterkfontein, South Africa

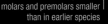

> 8 MYA   **PRESENT**   **AGE** *2.4–1.6 mya*
> 7 MYA   1 MYA   Dated using absolute dates from layers of volcanic ash and basalt above and below the fossils
> 6 MYA   2 MYA
> 5 MYA   3 MYA
> 4 MYA

> **FOSSIL RECORD** Several skulls and sections of crania, fragments of hand, arm, leg, and foot bones, and a partial skeleton

JAWS AND TEETH The lower face of *Homo habilis* is more slender than that of *Australopithecus* or *Paranthropus* species. The molars and premolars are narrow and reduced in overall size, which could suggest a diet requiring less chewing or perhaps better quality foods eaten in smaller quantities. The incisors, and especially the canines, are relatively large and the whole front tooth row is expanded. Although the palate is short, the region of bone holding the roots of the teeth is long and well separated from the nasal opening. The body of the lower jaw, or mandible, is less deep than in australopithecines, with a receding chin, and the base of the mandible is rounded.

relatively large incisors and canines, but small compared to australopithecines

molars and premolars smaller than in earlier species

fairly thin coating of enamel

teeth are arranged in a more rounded arc, like those of modern humans

OH8 FOOT
This specimen is one of the most complete feet known for any fossil hominin, missing only the toes and part of the heel bone. It belonged to a young *H. habilis* who suffered from arthritis after sustaining an injury to the foot.

OH7 JAW
Jaws are particularly useful finds, especially if, like the jaw of OH7, they retain many of their teeth. The relative importance of the back grinding teeth and the front shearing teeth can tell us a great deal about the diet of our ancestors, and suggests that *H. habilis* ate more meat than other primates.

FEET The foot of *H. habilis* is in many ways similar to that of modern humans in that it had limited mobility at most joints, short toes, alignment of the four smaller toes, and a moderate arch. It is likely that the big toe was not held as close to the others as it was in later *Homo* species.

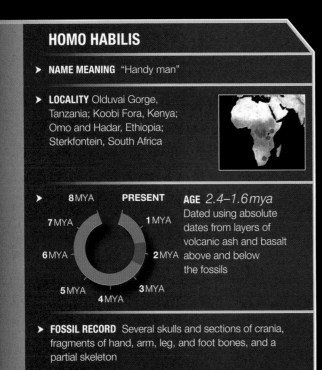

this foot specimen lacks a heel and toe bones, but the shape is clearly similar to a human foot

Physical features

Since the discovery of the first *Homo habilis* fossils from Olduvai, new specimens have been described from several other sites, including Koobi Fora in Kenya and the Omo River basin in Ethiopia. Together, these finds describe a species that was, generally speaking, relatively small in stature and more lightly built than most australopithecines, but which had a larger brain and a body capable of bipedal locomotion. The arms were longer and perhaps stronger than those of modern humans. However, there is enough variation in these traits across specimens to generate controversy. Some anthropologists have argued for the placing of these specimens within the genus *Australopithecus*, or for the splitting of the taxon into sub-groups.

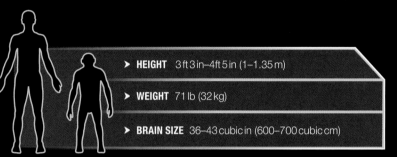

> **HEIGHT** 3 ft 3 in–4 ft 5 in (1–1.35 m)

> **WEIGHT** 71 lb (32 kg)

> **BRAIN SIZE** 36–43 cubic in (600–700 cubic cm)

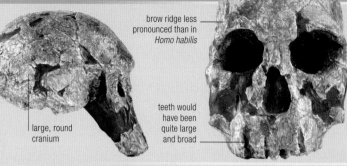

brow ridge less pronounced than in *Homo habilis*

large, round cranium

teeth would have been quite large and broad

In 1972, a team led by Richard Leakey working beside Lake Turkana, Kenya, found the fragmented but complete skull KNM-ER 1470. It was a large-brained hominin with a cranial capacity of 46–49 cubic in (750–800 cubic cm), a flat face, and forward-facing cheekbones. It was tentatively ascribed to the genus *Homo*, but Leakey was not confident that it was *H. habilis*. In 1992, anatomist Bernard Wood suggested that KNM-ER 1470 be placed in a new taxon, *H. rudolfensis*. In this scenario, there were two species of early *Homo* at 1.9 million years ago—one small and more primitive-bodied, and the other larger-bodied and larger-brained. No real consensus was reached, and the debate continues.

BODY AND LIMBS Few of the presumed *Homo habilis* skulls have been found with associated skeletons. The limited evidence suggests that the hand was wide, with a large thumb capable of a precision grip. The gait, although probably bipedal, may have been different to that from modern humans.

large, domed braincase

CRANIUM The cranium of *Homo habilis* is rounded and relatively gracile (lightly built), with an estimated cranial capacity of 36–43 cubic in (600–700 cubic cm). The frontal region is expanded compared to the australopithecines, and some specimens have marked neck-muscle attachment areas at the back of the skull.

KNM-ER 1813 CRANIUM
The KNM-ER 1813 cranium was found in relatively good condition. It is smaller than many of the other *H. habilis* finds but has stronger brow ridges than larger specimens.

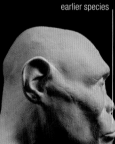

lower face projects forward less than in earlier species

expanded braincase compared to australopithecines

pronounced bone ridges over eyes

FACE Compared to the australopithecines, the face is small and lightly built. There is a continuous bony brow ridge above both orbits, which are widely spaced. A clear spine exists for the attachment of the nose septum, and the whole face projects slightly forward.

teeth are quite long and narrow

SMALL SKULL
Although quite similar to its australopithecine ancestors in other respects, *H. habilis*'s face looked more human because of its shorter jaw and smaller chewing muscles.

Archaeology

By 2.6 million years ago, simple flakes, cores, and hammerstones made up the Oldowan tool industry, first described by Mary Leakey at Olduvai Gorge. It was long thought that *Homo habilis* was the first hominin to make stone tools—of the Oldowan type—and that stone tool technology was a defining characteristic of early *Homo*. But even earlier stone tools from Lomekwi, Kenya, dating to around 3.3 million years ago, and marks found on animal bones from Dikika, Ethiopia, predate our genus. Other species that may have made stone tools include *Kenyanthropus platyops*, australopithecines, and paranthropines. Of course, tools made from perishable materials such as wood or leaves may have been used much earlier.

Oldowan tools

Louis and Mary Leakey (see p.94) were initially drawn to work at Olduvai Gorge by the richness of its archaeology—in particular, discoveries of stone tools made of basalt, quartz, and quartzite that were derived from geological deposits dating from between 2.2 and 1.7 million years ago. Many thousands of artifacts were collected through surface survey and excavation. Mary Leakey painstakingly analyzed them, and she was the first to describe and name the tool types, which together became known as the Oldowan tool industry. This simple, effective stone-tool technology was used for at least 700,000 years across sub-Saharan Africa.

STONE KNAPPING
A chopper can be made by striking flakes from a rounded cobble using another stone, to give a sharp cutting edge. This requires learning and good hand/eye coordination. Tools made in this way may have been used to butcher carcasses or chop plant material.

non-cutting side for gripping

flaked surface

FLAKE

FLAKED COBBLE

smooth handle surface

tool made of quartzite

CHOPPERS

TOOL TYPES
The earliest Oldowan tools are mainly made from cobbles of rough stone such as basalt, quartz, and quartzite, struck with a hammerstone to produce cores and sharp flakes for cutting.

Olduvai Gorge

A steep-sided, double-branching ravine in the Serengeti Plains of Tanzania, the Olduvai Gorge is approximately 30 miles (48 km) long and 300 ft (90 m) deep. First described by western explorers in 1911, the region became known for its intriguing stone-tool archaeology and ancient mammalian fauna, and in 1931 Louis Leakey embarked on his first expedition to the region.

The geology of Olduvai Gorge is complex, being formed largely of sedimentary lake deposits, lava flows, and volcanic-ash deposits that, over time, have experienced fault shifts and been subjected to erosion. Because these volcanic layers are datable (see p.10), the whole sequence of deposits at Olduvai has proved invaluable to archaeology. The site is divided into seven major "Beds," the oldest of which is Bed I, from which the *Paranthropus boisei* "Zinj" skull OH5 was excavated in 1959. The fragmented remains of more than 50 hominins are now known from these Beds, dated between 1.75 million to 15,000 years ago, making it one of the most continuous records of human evolution known.

SURVEYING THE BEDS
Louis Leakey (right) examines exposed evidence of a hominin campsite among the clay and ash layers of Olduvai's Bed I, 1962.

OLDUVAI BEDS
Largely identified through the work of geologist Richard Hay, the distinct layers, or Beds, at Olduvai record periods of deposition, erosion, and landscape change. In addition to Beds I–IV, there are also three later beds: the Masek, Ndutu, and Naisiusiu. A complete *Homo sapiens* skeleton was discovered in the Naisiusiu Bed.

	GEOLOGY	FOSSIL RECORD
BED IV	Divided into a lower layer of clays, sandstones, and conglomerates, and an upper layer of volcanic-ash deposits.	Animal fossils, later Acheulian tools, and evidence of *Homo erectus*, but hominin finds are rarer than in Beds I and II.
BED III	Stream and wind-blown deposits laid down after earth movements and an arid climate had removed the lake.	Animal fossils, later Acheulian tools, and evidence of *Homo erectus*, but hominin finds are rarer than in Beds I and II.
BED II	Lava flows, volcanic-ash deposits, and other sediments laid down on the floor of a lake basin.	Animal fossils, remains of *Paranthropus boisei*, *Homo habilis*, and *H. erectus*, later Oldowan and early Acheulian tools.
BED I	Lava flows, volcanic-ash deposits, and other sediments laid down on the floor of a lake basin.	Animal fossils, remains of *Paranthropus boisei* and *Homo habilis*, crude Oldowan pebble tools, and evidence of campsites.

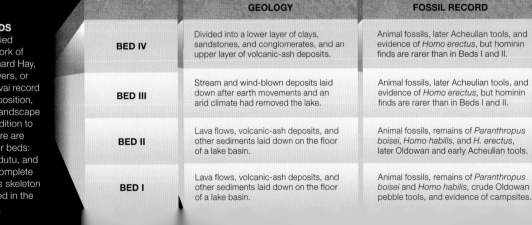

Archaeological assemblages

An assemblage is a group of materials found in association with each other that represents the result of human activity. In the Oldowan, the character of assemblages varies across sites: some have very dense accumulations of stone tools, while others have a sparse distribution; and some sites are dominated by a particular tool type or show evidence of weathering. Such details can be used to interpret the site, perhaps indicating how long the site was occupied or what activities took place there.

Many Oldowan sites contain assemblages of both stone tools and animal skeletons, and interpreting the relationship between the two is rife with controversy. For example, the Olduvai Gorge site called FLK North 6 included an elephant skeleton. The elephant was nearly complete, except for the tusks and skull, but the bones were displaced. Over 100 stone tools were discovered near the body, and various anthropologists have suggested different interpretations for this site. In 1971, Mary Leakey described it as a butchering site where hominins processed their kill. Since then, others have reported cutmarks and carnivore gnawmarks on the bones, suggesting that the hominins were scavengers. Recently it has been proposed that hominins may not have been processing the elephant remains at this site at all.

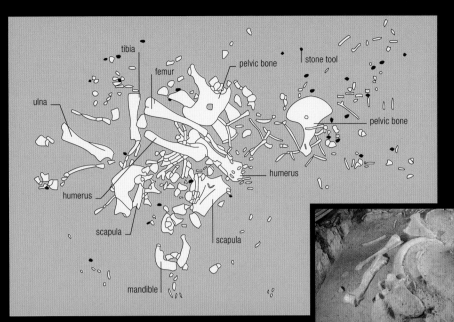

SURFACE PLAN
As an excavation proceeds, a meticulous record is made of the exact positions in which objects are found. This site plan, based on one drawn up by Mary Leakey, shows the elephant skeleton at Olduvai's FLK North 6 site surrounded by other animal bones and stone-tool debris (the black dots).

ELEPHANT BONES
The elephant found at Olduvai was *Elephas recki*, a large, extinct relative of the Asian elephant. This collection of *E. recki* bones is preserved in Kenya's Sibiloi National Park.

REVEALING LANDSCAPE
Once a lake environment, Olduvai Gorge was created when the lake basin was uplifted and cut by streams. Erosion and earth movements have exposed many rock layers containing fossils and human artifacts that reveal much about the past.

Homo habilis:

A group gathers around the carcass of a rhinoceros, butchering the meat using Oldowan stone tools.

HUNTER OR SCAVENGER?

Whether *Homo habilis* actively hunted animals or scavenged the prey of large carnivores is uncertain. Some scientists view hunting as an important activity in the context of early hominin evolution in that it may require cooperation, planning ahead, and physical skill. Whether *H. habilis* possessed all these abilities is a contentious issue.

LAKE SEDIMENTS

Homo habilis fossils excavated in the Olduvai Gorge, in Tanzania, come from soil layers made of sediment from a small lake that had formed in the area by about 2 million years ago. Its water level fluctuated greatly with the seasons, sometimes leaving flooded grassland around its margins.

CORE CHOPPER
Homo habilis would probably have used a stone chopper to dismember a carcass, then crush the animal's bones for the marrow within them. A chopper would have been produced when flakes were detached from a chunk of rock to leave a core. Although crude, they would have had sharp edges, which made them versatile tools.

BUTCHERING A CARCASS
Sharp flakes—produced by a hammerstone striking a core—would have been ideal for removing an animal's skin, scraping the hide, and slicing the meat. Some animal bones found in fossil beds associated with *Homo habilis* are thought to bear cut marks produced by using tools to strip meat from the bones.

MAKING TOOLS
Until recently, *Homo habilis* was widely believed to have been the first hominin to use stone tools. But tools and animal bones thought to have been smashed open now predate the appearance of this species (see p.32). However, the Oldowan tools frequently associated with *Homo habilis* are more sophisticated and widespread.

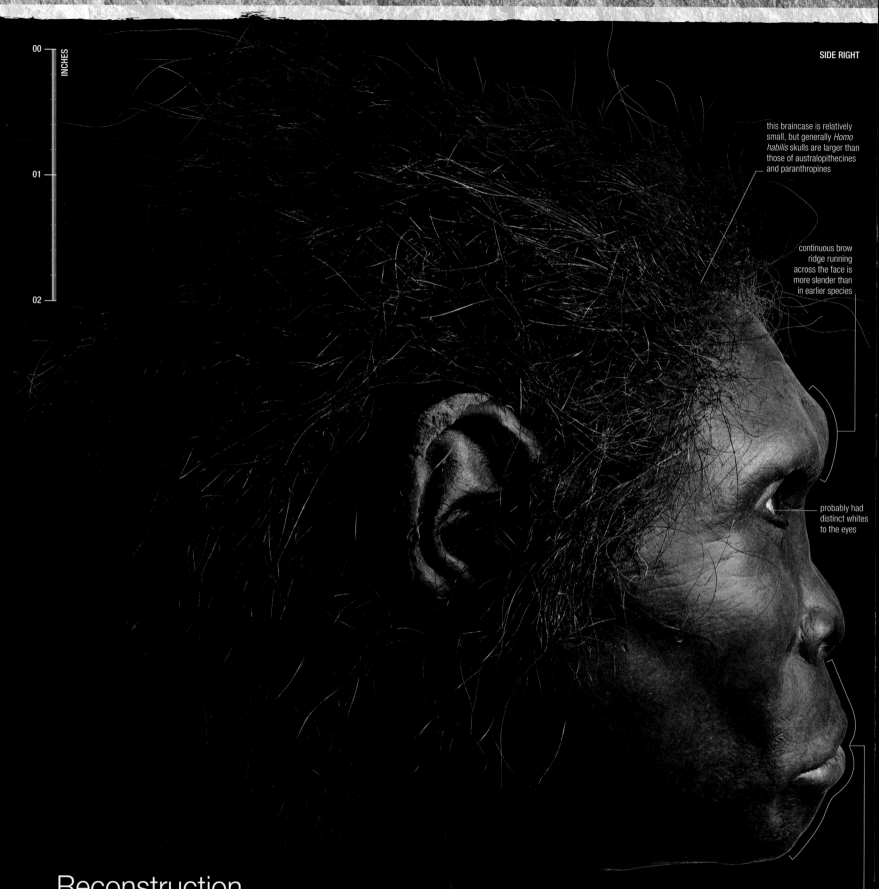

this braincase is relatively small, but generally *Homo habilis* skulls are larger than those of australopithecines and paranthropines

continuous brow ridge running across the face is more slender than in earlier species

probably had distinct whites to the eyes

lower face projects forward less than in australopithecines

Reconstruction

This adult female is modeled on the KNM-ER 1813 skull from Koobi Fora, Kenya—one of the most complete *Homo habilis* skulls known, and dated to about 1.85 million years ago. It is the smallest of several skulls found at Lake Turkana, and its discovery prompted some anthropologists to divide the fossil material from this period into two separate species: *Homo habilis* and *Homo rudolfensis*. Because this skull was not associated with a mandible, the modelers have used the OH13 mandible from Olduvai Gorge in Tanzania, dated to around 1.7 million years ago, as a basis for reconstructing her jaw.

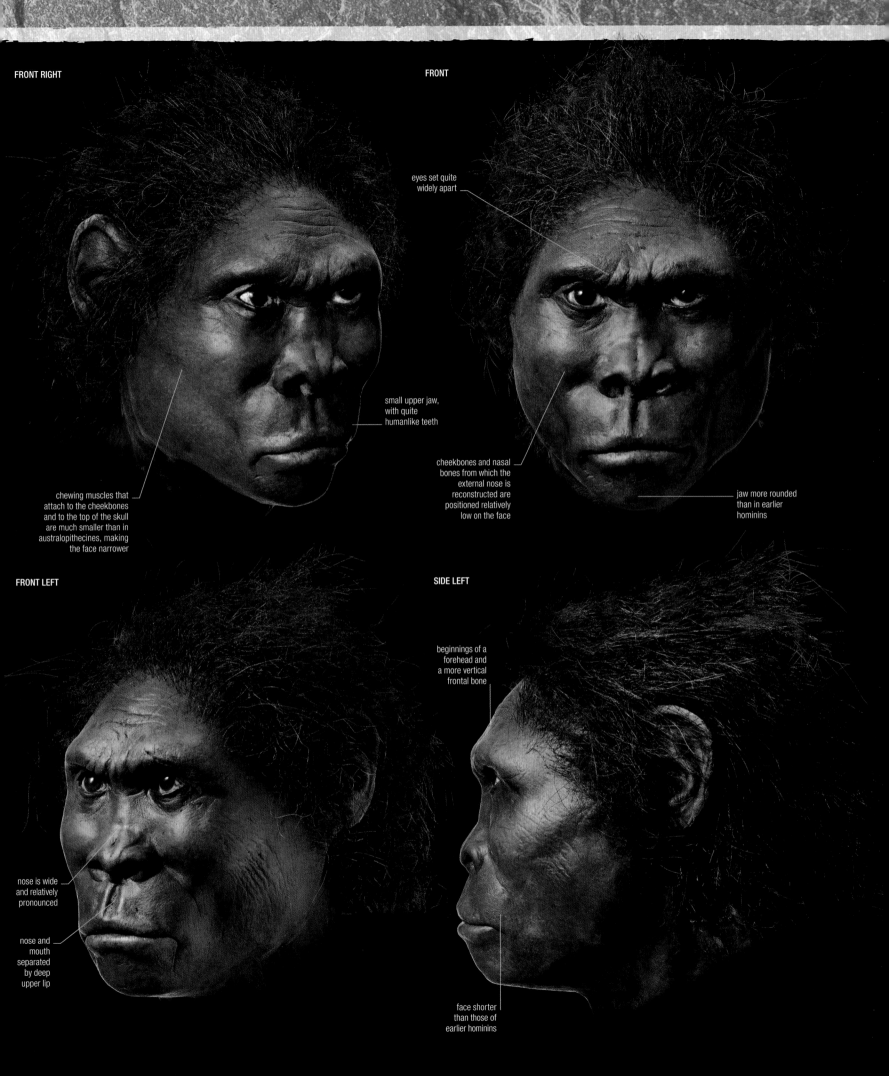

FRONT RIGHT

FRONT

eyes set quite
widely apart

small upper jaw,
with quite
humanlike teeth

cheekbones and nasal
bones from which the
external nose is
reconstructed are
positioned relatively
low on the face

jaw more rounded
than in earlier
hominins

chewing muscles that
attach to the cheekbones
and to the top of the skull
are much smaller than in
australopithecines, making
the face narrower

FRONT LEFT

SIDE LEFT

beginnings of a
forehead and
a more vertical
frontal bone

nose is wide
and relatively
pronounced

nose and
mouth
separated
by deep
upper lip

face shorter
than those of
earlier hominins

Homo *georgicus*

Represented by a collection of fossils from
a single site in Georgia, *Homo georgicus* was
one of the first hominins to leave Africa.

SKULL IN-SITU
Archaeologists examine a Dmanisi skull that was found in ancient deposits full of the fossils of other extinct animals, such as deer, wolves, and large cats. The good state of preservation suggests the bodies were covered soon after death.

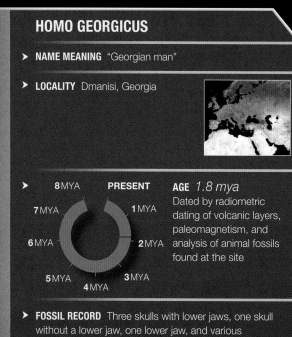

HOMO GEORGICUS

> **NAME MEANING** "Georgian man"

> **LOCALITY** Dmanisi, Georgia

>

8 MYA	PRESENT	**AGE** *1.8 mya*
7 MYA	1 MYA	Dated by radiometric dating of volcanic layers, paleomagnetism, and analysis of animal fossils found at the site
6 MYA	2 MYA	
5 MYA	3 MYA	
4 MYA		

> **FOSSIL RECORD** Three skulls with lower jaws, one skull without a lower jaw, one lower jaw, and various fragments from the upper and lower body, including arm, leg, hand, and foot bones

Homo georgicus is controversial. The fossils from Dmanisi constitute the earliest well-dated Eurasian population of hominin. Whether this represents a new species is unclear, but the find is of great importance to our understanding of early hominin migrations.

Discovery

The paleoanthropological site of Dmanisi is located under the ruins of a medieval town, about 53 miles (85 km) southwest of Tbilisi, in Georgia. In 1983, medieval archaeologists working there found the bones of extinct animals. The discovery of stone tools confirmed the suspicion that the ruins overlay a more ancient site, and excavation of the 1.7–1.8 million-year-old deposits revealed a hominin jawbone in 1991. Since then fossils representing males, females, and juveniles have been collected. Initially assumed to represent a European population of *Homo erectus*, in 2002 the material was described as a new species, *Homo georgicus*, with a large jawbone (D2600) presented as the type specimen. In 2006, another reassessment placed most of the material back into *H. erectus*. This controversy is partly fueled by the variability in body sizes, coupled with the light build of the skulls in comparison to Asian *H. erectus*.

FOSSIL BOOTY
At the Dmanisi site, anthropologists have the unusual experience of excavating within the walls of a ruined medieval site that lies above the much older fossil deposits.

Physical features

The Dmanisi fossils provide an important record of the earliest hominin populations to leave Africa. Including male and female skulls, many teeth, and bones from across the body, the collection gives fascinating insight into the biology, health, and lifestyles of these people. Stature and body mass has been calculated from measurements of the arm and leg bones, and they were smaller than most later hominins. Their limb proportions were similar to modern humans and the earliest African *Homo*, but they had modest cranial capacity, which overlaps with that of *H. habilis*, but would be below average for *H. erectus*.

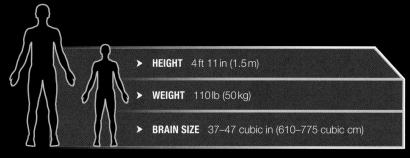

> **HEIGHT** 4 ft 11 in (1.5 m)
>
> **WEIGHT** 110 lb (50 kg)
>
> **BRAIN SIZE** 37–47 cubic in (610–775 cubic cm)

UPPER BODY Several vertebrae have been found, representing the neck, thorax, and lower parts of the spine. They all have relatively large joint surface areas and proportions more similar to *Homo erectus* than australopithecines, suggesting a modern type of spinal curve. In contrast, the Dmanisi hominins had a more australopithecine-like upper limb morphology (form and structure), with a low degree of torsion (twist) in the humerus (upper arm bone).

shape of vertebrae suggests curve of lower back already present

vertebra

clavicle (collarbone)

upper rib

humerus lacks slight "twist" that this bone has in modern humans

lower rib

A SMALL FEMALE
One of the partial skeletons is likely to be that of a subadult female. Some of the bones show incomplete growth, and may be associated with a gracile (lightly built) skull whose "wisdom" teeth were not fully erupted.

humerus

SMALL BRAINS
The average brain size of specimens from Dmanisi is quite low, close to that of *Homo habilis*, but the shape of the face and jaw display similarities with some *Homo erectus* specimens.

smalll skull

no external bony nose

SKULL The skulls are relatively small, slightly elongated, and with a thickened keel that runs from front to back along the midline of the braincase. The base of the skull is wide, and the back is smooth, rounded, and quite gracile. The brow ridge runs above the two eye sockets, but is only moderately robust. The upper face is quite small, with narrow nasal bones, but the upper jaw region is mildly prognathic (projecting forward).

kneecap is less symmetrical than in modern humans

long and straight femur, as in modern humans, but angle of the neck is intermediate between that of humans and australopithecines

shin bone shows that foot was habitually flexed against lower leg

SUPPORTIVE CARE

One of the best preserved and most intriguing skulls from Dmanisi is that of an elderly individual who had lived for some time with no teeth at all. When teeth are lost in life the empty sockets within the bone resorb and create a characteristically scooped-out shape to the jaw. To survive for so long, this person must have lived within a supportive social group.

LOWER BODY All the bones of the leg and many of the foot are represented in the collection of fossils. The femur (thigh bone) is angled under the body to ensure good balance. The length of the combined bones of the hind limb is essentially modern and the foot, which had a big toe that lay parallel to the other toes, also suggests that the mode of walking was probably highly efficient, similar to that of modern humans. There are a few clues, however, that bipedalism in the Dmanisi population was not identical to ours. One example is found in the foot, which is orientated medially (turned inward), indicating more equal load distribution than is seen in modern human walking.

metatarsals (foot bones) suggest that foot had arched structure

FOOT ARCHES
The metatarsal bones form transverse and longitudinal arches that are very similar to those of modern humans. These are essential to efficient bipedal walking because they are weight bearing and allow good propulsion at the toe push-off.

tip of toe

talus (one of the ankle bones) suggests that *H. georgicus*'s feet turned inward

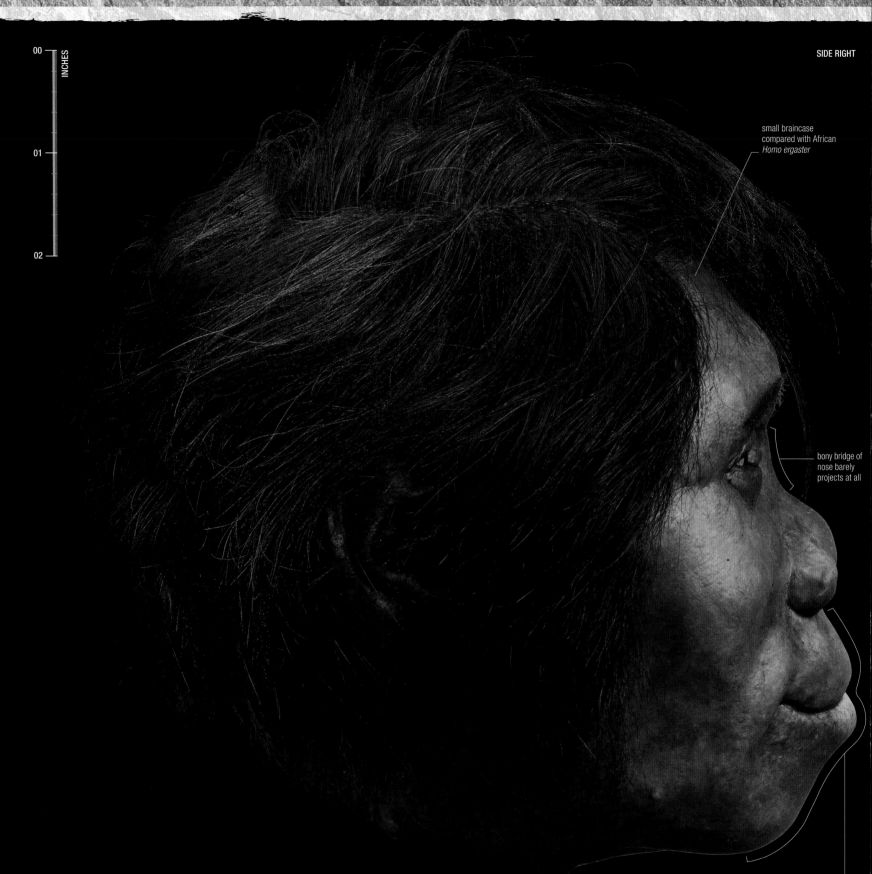

small braincase
compared with African
Homo ergaster

bony bridge of
nose barely
projects at all

jaw projects quite far
forward from face

Reconstruction

This representation of *Homo georgicus* is based on the skull D2700, whose
small size and relatively delicate features suggest that it may have been a
female. D2700 was one of five skulls found at Dmanisi, Georgia, between
1991 and 2005. Dated to around 1.8 million years ago, Dmanisi is the earliest
known hominin site outside of Africa. Although the fossils have been placed in
their own species, many anthropologists think they are a geographical variant
of *Homo ergaster* and/or *Homo erectus*. Some argue that D2700 looks like the
skull KNM-ER 1813, currently assigned to *Homo habilis* (see pp.106–07).

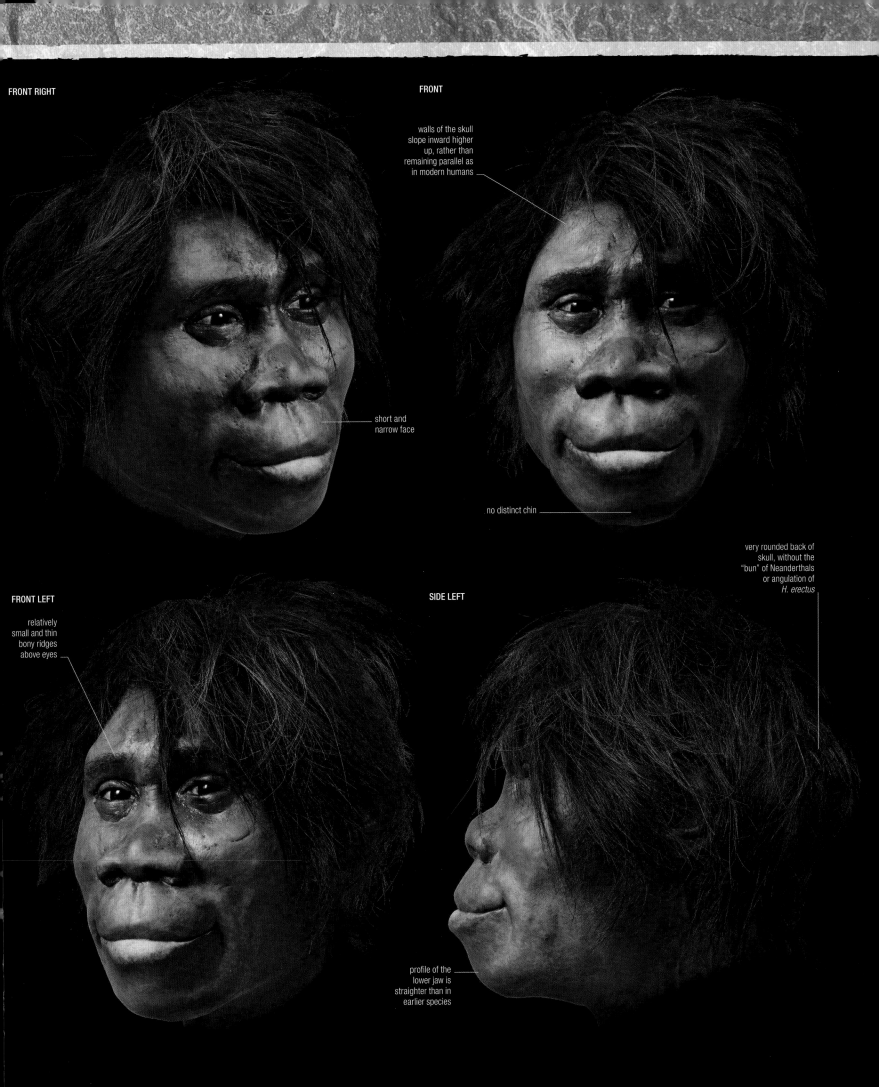

walls of the skull
slope inward higher
up, rather than
remaining parallel as
in modern humans

short and
narrow face

no distinct chin

very rounded back of
skull, without the
"bun" of Neanderthals
or angulation of
H. erectus

relatively
small and thin
bony ridges
above eyes

profile of the
lower jaw is
straighter than in
earlier species

HOMO GEORGICUS | 113

Homo ergaster

As tall as modern humans, and with a similar build, *Homo ergaster* would have appeared very different to its apelike ancestors.

Homo ergaster, or the "workman," after the rich stone-tool assemblages sometimes associated with this species, was the first hominin to show a modern humanlike body shape, stature, and limb proportions. This species probably had a mode of walking and running very similar to our own.

Discovery

The first specimen to be described as *Homo ergaster* was a well-preserved lower jaw (KNM-ER 992) found in 1971 at Ileret in East Turkana, Kenya. Other fossil skulls of both sexes have since been identified, but by far the most striking specimen is KNM-ER WT 15000, the remarkably complete skeleton of an adolescent boy, which is also known as "Turkana Boy" or "Nariokotome Boy." Discovered in 1984 by Kenyan achaeologist Kamoya Kimeu at West Turkana, and described by a collaborative team of researchers lead by British anthropologist Alan Walker and Kenyan anthropologist Richard Leakey, this 1.5 million-year-old specimen was placed in *Homo ergaster* by British paleoanthropologist Bernard Wood in 1992. The skeleton lacks only the left humerus (upper arm bone), the radius and ulna (lower arm bones), and the bones of the hands and feet, and therefore gives scientists unprecedented insight into the body proportions, biomechanics, and growth of this hominin species.

FOSSIL HUNTERS
Kamyoya Kimeu (left) was one of the team led by Richard Leakey (right) working at Lake Turkana, Kenya, when he discovered Turkana Boy, the best known of all the *Homo ergaster* specimens.

KOOBI FORA RESEARCH CENTRE
The research centre at Koobi Fora is on the opposite side of Lake Turkana in Kenya from where Turkana Boy was found. Koobi Fora is a rich fossil site, and is now part of Sibiloi National Park, a UNESCO World Heritage site.

HOMO ERGASTER

> **NAME MEANING** "Workman"

> **LOCALITY** Various sites in the East African Rift Valley—Kenya, Tanzania, and Ethiopia—as well as South Africa

> 8 MYA   **PRESENT**   **AGE** *1.9–1.5 mya*
> 7 MYA   1 MYA   Mainly dated through absolute dating of volcanic ash layers above and below the fossils
> 6 MYA   2 MYA
> 5 MYA   3 MYA
> 4 MYA

> **FOSSIL RECORD** One nearly complete skeleton and a few complete skulls; also various fragments of skull, jaw, pelvis, and limb bones

OLDEST EVIDENCE OF MODERN HUMANLIKE FEET

Several sets of hominin footprint trails, thought to have been made by either *Homo ergaster* or *Homo erectus*, have been found in the Koobi Fora formation, near Ileret, in Kenya. Laser-scanning and analysis of these prints have revealed evidence for a modern humanlike foot anatomy as early as 1.51 million years ago. The prints demonstrate that the walkers had big toes that were relatively in-line rather than separated from the other toes. The prints also show that the walkers had well-defined arches to their feet and the typical pattern of weight distribution—shifting from the heel to the ball of the foot—seen in modern human walking.

print of modern-shaped foot   weight has been transferred to ball of foot

JAWS AND TEETH The face projects forward, but the lower jaw recedes and does not have an external chin. The premolar and molar teeth are much smaller relative to the front teeth than is found in australopithecines. The roots of the upper front teeth are vertical. On Turkana Boy's jaw, where a milk tooth has been lost, there is evidence of infection, and it has been suggested he could have died from the resulting blood poisoning.

UPPER BODY The arms are short relative to the legs, and the torso slender and barrel shaped, much more similar to modern humans than earlier hominins. Slight differences in the shoulder blade compared to modern humans may suggest a different use of the upper limbs, perhaps related to a long period of infant crawling. Some of the vertebrae show a restricted space for the spinal cord, which might mean the individual did not have the breathing control necessary for modern speech.

REMARKABLE FOSSIL
Although few *Homo ergaster* fossils have been discovered, Turkana Boy is so complete that we can deduce a great deal about what his species looked like and how it lived. *H. ergaster* was the first hominin to achieve the stature of modern humans. This individual was quite tall—although he was young, he stood about 5 ft 2 in (1.6 m) tall and may have grown to be up to 5 ft 5 in (1.65 m).

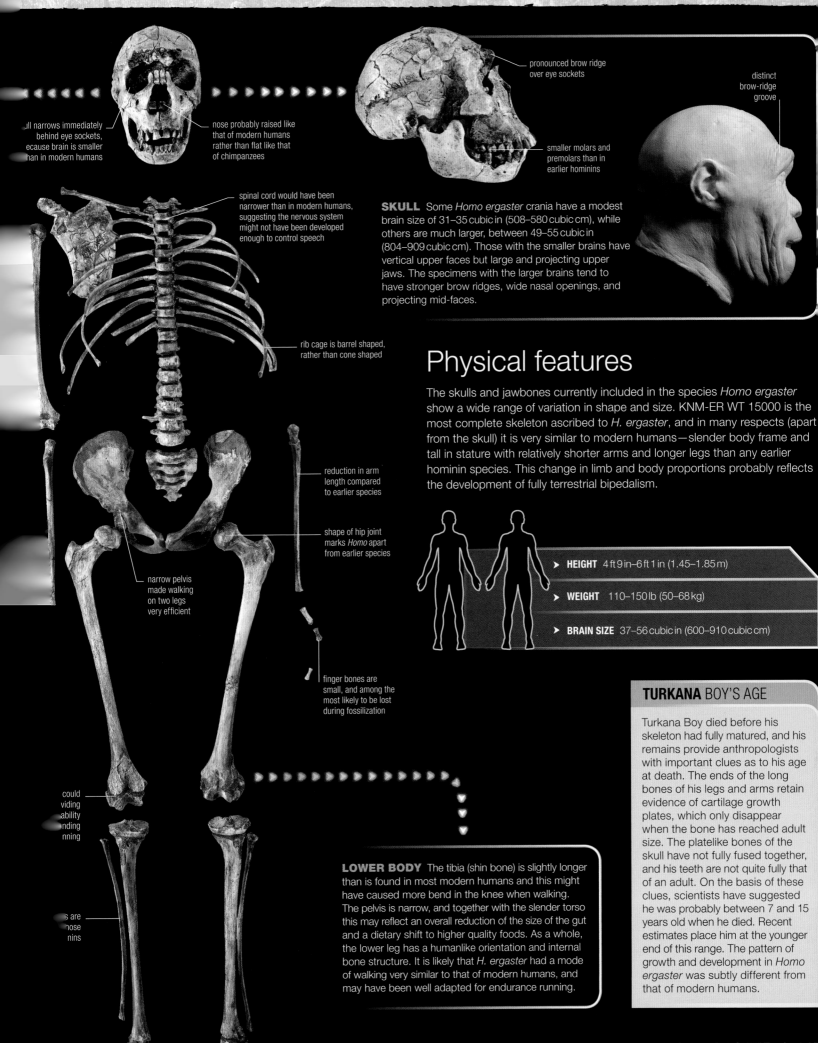

pronounced brow ridge
over eye sockets

ull narrows immediately
behind eye sockets,
because brain is smaller
than in modern humans

nose probably raised like
that of modern humans
rather than flat like that
of chimpanzees

smaller molars and
premolars than in
earlier hominins

distinct
brow-ridge
groove

spinal cord would have been
narrower than in modern humans,
suggesting the nervous system
might not have been developed
enough to control speech

SKULL Some *Homo ergaster* crania have a modest
brain size of 31–35 cubic in (508–580 cubic cm), while
others are much larger, between 49–55 cubic in
(804–909 cubic cm). Those with the smaller brains have
vertical upper faces but large and projecting upper
jaws. The specimens with the larger brains tend to
have stronger brow ridges, wide nasal openings, and
projecting mid-faces.

rib cage is barrel shaped,
rather than cone shaped

Physical features

The skulls and jawbones currently included in the species *Homo ergaster*
show a wide range of variation in shape and size. KNM-ER WT 15000 is the
most complete skeleton ascribed to *H. ergaster*, and in many respects (apart
from the skull) it is very similar to modern humans—slender body frame and
tall in stature with relatively shorter arms and longer legs than any earlier
hominin species. This change in limb and body proportions probably reflects
the development of fully terrestrial bipedalism.

reduction in arm
length compared
to earlier species

shape of hip joint
marks *Homo* apart
from earlier species

narrow pelvis
made walking
on two legs
very efficient

> **HEIGHT** 4 ft 9 in–6 ft 1 in (1.45–1.85 m)

> **WEIGHT** 110–150 lb (50–68 kg)

> **BRAIN SIZE** 37–56 cubic in (600–910 cubic cm)

finger bones are
small, and among the
most likely to be lost
during fossilization

could
viding
ability
nding
nning

TURKANA BOY'S AGE

Turkana Boy died before his
skeleton had fully matured, and his
remains provide anthropologists
with important clues as to his age
at death. The ends of the long
bones of his legs and arms retain
evidence of cartilage growth
plates, which only disappear
when the bone has reached adult
size. The platelike bones of the
skull have not fully fused together,
and his teeth are not quite fully that
of an adult. On the basis of these
clues, scientists have suggested
he was probably between 7 and 15
years old when he died. Recent
estimates place him at the younger
end of this range. The pattern of
growth and development in *Homo
ergaster* was subtly different from
that of modern humans.

s are
nose
nins

LOWER BODY The tibia (shin bone) is slightly longer
than is found in most modern humans and this might
have caused more bend in the knee when walking.
The pelvis is narrow, and together with the slender torso
this may reflect an overall reduction of the size of the gut
and a dietary shift to higher quality foods. As a whole,
the lower leg has a humanlike orientation and internal
bone structure. It is likely that *H. ergaster* had a mode
of walking very similar to that of modern humans, and
may have been well adapted for endurance running.

Archaeology

About 1.65 million years ago in eastern and southern Africa, and often associated with *Homo ergaster* or *Homo erectus*, a new tool form appears in the archaeological record. This is known as the Acheulean stone-tool industry. Like the Oldowan stone-tool industry (see p.102), it includes tool types such as flakes and choppers, but with the addition of an innovative new form—the handax.

Acheulean tools

Often large and made from stone with good fracturing properties, Acheulean handaxes were created through the bifacial knapping of a nodule. This involves striking a flake off one side, and then turning the nodule over to knap the other side, generating a fairly straight and sharp edge. It is a process requiring skill, forward planning, and the ability to imagine and then produce the standardized form. Early handaxes, such as those that have been found at Konso-Gardula in Ethiopia and Olduvai Gorge in Tanzania, tend to be simply worked pointed bifaces created with relatively few knapping strikes. In later examples, the variety of bifaced tools broadens to include cleaver, oval, and highly symmetric forms. There is debate over the function of these handaxes. Some experimental and microwear studies have suggested they were hunting weapons or butchery tools, others that they were wood choppers. Whilst the handaxe itself may have been used as a heavy-duty tool, the flakes taken from it would also have been useful cutting tools.

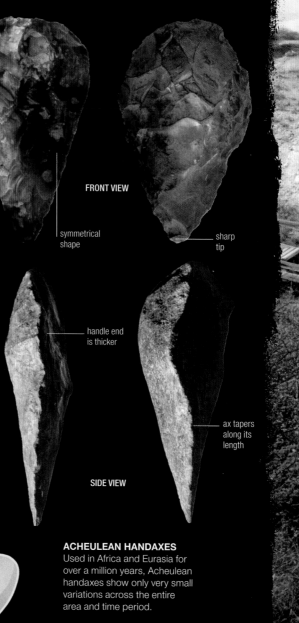

FRONT VIEW

symmetrical shape

sharp tip

handle end is thicker

ax tapers along its length

SIDE VIEW

ACHEULEAN HANDAXES
Used in Africa and Eurasia for over a million years, Acheulean handaxes show only very small variations across the entire area and time period.

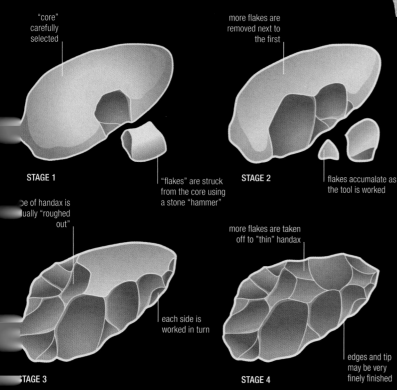

"core" carefully selected

STAGE 1

more flakes are removed next to the first

"flakes" are struck from the core using a stone "hammer"

STAGE 2

flakes accumalate as the tool is worked

be of handax is ually "roughed out"

STAGE 3

each side is worked in turn

more flakes are taken off to "thin" handax

STAGE 4

edges and tip may be very finely finished

MAKING A HANDAX
t takes considerably more skill to make an Acheulean handax than a Oldowan tool. A knapper would have o think several steps ahead when selecting a suitable piece of stone, and then prepare and place each strike precisely. Many handaxes were finished to a very high evel of symmetry, sometimes using several different ools, including stone hammers.

LANGUAGE AND TOOLMAKING

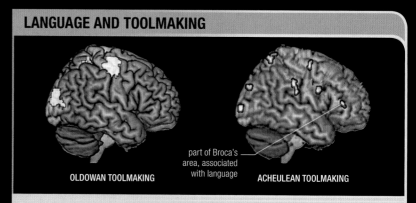

OLDOWAN TOOLMAKING

part of Broca's area, associated with language

ACHEULEAN TOOLMAKING

Archaeologists have long considered the manufacture of handaxes to be a skilled process requiring complex cognitive abilities. Many efforts have been made to teach other primates to knap stone tools, but with unremarkable results. Recently neuroscientists recorded the brain activity of human subjects while they knapped Oldowan and Acheulean tools. PET (positron emission tomography) scans demonstrated a high degree of activation in the premotor cortex when making Acheulean tools—the same region that is involved in understanding language. It may be that language and toolmaking have a shared evolutionary history, reinforcing each other through time.

OLORGESAILIE TOOL "FACTORY"
Hundreds of Acheulean handaxes and
other stone tools lie scattered across the
ground at Olorgesailie, Kenya, about
43 miles (70km) from Nairobi. The site
shows evidence of consistent human
activity from 1.2 to 0.4 million years ago.

Handax assemblages

Some very rich handax assemblages are known in the African Acheulean.
At Olorgesailie, Kenya, thousands of stone tools have been discovered as
surface finds or through excavations, which began in 1942 under the
direction of Mary and Louis Leakey. A considerable number of these tools
are handaxes, and while some of them are broken and may have been
discarded by their makers, many are intact. Archaeologists have discovered
that even during the early period of activity at Olorgesailie, around 1.2 million
years ago, hominins were using at least 14 different types of volcanic rock
as raw material for their tools, all of which were sourced from local quarries.
Thousands of waste stone flakes have also been found at these sites,
showing that these were the toolmaking locations. Some types of rock, such
as obsidian, are rare, and the nearest obsidian outcrop to Olorgesailie is
11 miles (18km) away from the main site. This gives a revealing insight into
the range of movement of these early toolmaking hominins.

Recent excavations at Olorgesailie have revealed assemblages not only
of stone tools but also of the remains of animal skeletons. One such site
includes the skeleton of an extinct elephant, *Elephas recki*, and several
thousand associated stone artifacts. Cut marks on the ribs and spine, and
the patterning of the flakes around the body, suggest that the tools were
manufactured, and the carcass butchered, right there.

EXCAVATION SITE
The site of the original excavations at Olorgesailie is open to
the public. The Olorgesailie assemblage is the largest known
concentration of handaxes. The handaxes, stone-tool debris,
and animal fossils found at this site were deposited on the
shifting shoreline of a now extinct lake. Some of the axes are too
large to be of obvious practical use, and are apparently unused.

head hair may have helped
prevent overheating in
hot climates

distinc
betwee
ridges
forehe

lower face projects
forward less than in
earlier species

Reconstruction

This adult female has been reconstructed from the KNM-ER 3733 cranium
found at Koobi Fora, Kenya, in 1975 and dated to 1.8–1.7 million years ago.
The skull was excavated from the same layer as one found six years previously
and now assigned to *Paranthropus boisei*. KNM-ER 3733 surprised many
anthropologists; at the time it was thought that only one hominin species would
exist at any time, since it is rare for two species competing for the same
ecological resources to coexist. The discovery of this *Homo ergaster* skull
was a key find in demolishing the "single species hypothesis."

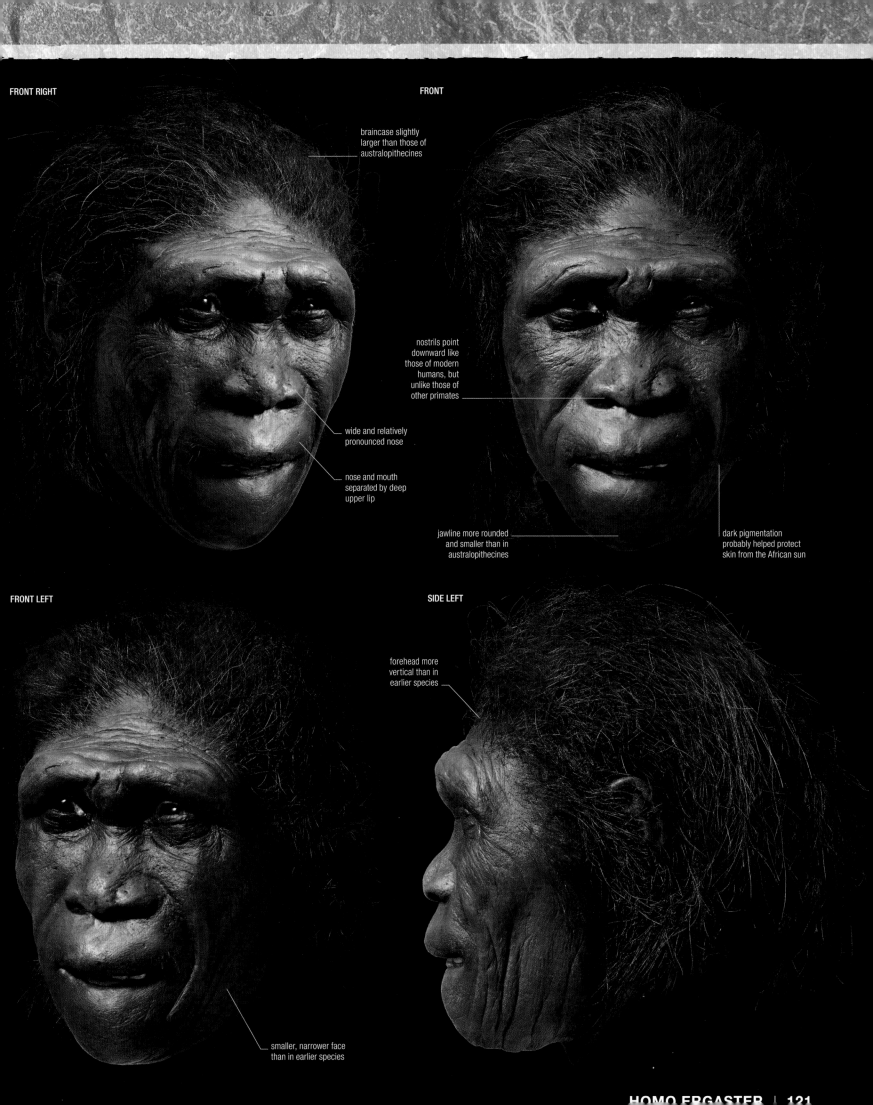

FRONT RIGHT

FRONT

braincase slightly
larger than those of
australopithecines

nostrils point
downward like
those of modern
humans, but
unlike those of
other primates

wide and relatively
pronounced nose

nose and mouth
separated by deep
upper lip

jawline more rounded
and smaller than in
australopithecines

dark pigmentation
probably helped protect
skin from the African sun

FRONT LEFT

SIDE LEFT

forehead more
vertical than in
earlier species

smaller, narrower face
than in earlier species

Homo erectus

While the origins of this species are unclear, *Homo erectus* is well represented by fossils in Asia, where it may have survived until 30,000 years ago.

Homo erectus was the first hominin species to be identified outside Europe. With a type specimen from Java, Indonesia, it is known from many sites across Asia. There is considerable disagreement as to whether European or African fossils should be assigned to this species. Here, the Asian fossils are presented as the core record.

Discovery

The first *Homo erectus* fossil, initially known as *Anthropopithecus erectus*, was found by Eugène Dubois on the island of Java in 1891, making the species one of the earliest ancient hominins to be discovered. In the 1920s, Davidson Black, Professor of Anatomy at Peking Union Medical College, identified some teeth discovered at the Chinese cave site Zhoukoudian as *Sinanthropus pekinensis*. By 1937, Black's team (he died in 1934) had unearthed a rich collection of fossils from the site, including facial fragments, jaws, and limb bones. Even as the prospect of war loomed, excavations continued. Paleontologist Ralph von Koenigswald found more specimens in Java, near Sangiran and Modjokerto. In 1939, von Koenigswald traveled to Beijing to compare his finds with those made at Zhoukoudian by Black and anthropologist Franz Weidenreich. Von Koenigswald and Weidenreich agreed that the fossils should be considered the same species. Most anthropologists now agree that these Asian fossils are correctly assigned to *H. erectus*. But controversy surrounds whether this species is also found in Africa and Europe, and how it is related to *H. ergaster* and *H. heidelbergensis*.

ZHOUKOUDIAN CAVES
These limestone caves, 26 miles (42 km) southwest of Beijing, have revealed some of the most important evidence of *Homo erectus* in Asia.

EUGENE **DUBOIS**

In 1887, the anatomist Eugène Dubois (1858–1940) joined the Royal Dutch East Indies Army to serve as a medical officer, but his agenda was clear—to find fossils of early humans in Asia. Having secured funds from the colonial government for his excavations, he began work near Trinil in the Solo River Valley, Java, in 1891. He was soon rewarded with the find of a partial skull, a tooth, and a humanlike thigh bone. Dubois proposed that they came from an ancient human *Anthropopithecus erectus*, which he later renamed *Pithecanthropus erectus*.

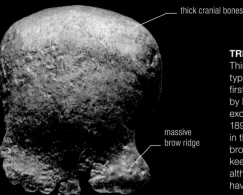

thick cranial bones

massive brow ridge

TRINIL 2
This skull cap—the species' type specimen—was among the first *Homo erectus* fossils found by Eugène Dubois during his excavations at Trinil, Java, in 1891. The skull has thick bones in the cranial vault, a prominent brow ridge, and a marked sagittal keel running along the midline, although many of its features have been worn flat with age.

HOMO ERECTUS

> **NAME MEANING** "Upright Man"

> **LOCALITY** Various sites in China and Java in Indonesia; African and European sites are contested

> 8 MYA   **PRESENT**   **AGE** *1.8 mya–30,000 ya*
> 7 MYA   1 MYA   Mainly dated through relative dating based on matching fossils found in caves with fossils from absolutely dated sites in East Africa
> 6 MYA   2 MYA
> 5 MYA   3 MYA
> 4 MYA

> **FOSSIL RECORD** Only one relatively complete cranium, several incomplete crania, as well as teeth and jaws, and a few limb bones

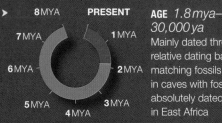

SOLO RIVER VALLEY
Most of the Javan *Homo erectus* fossils have been found in this valley. At one of the sites, Sangiran, the dating of volcanic sediments suggests that the fossils are at least 1.5 million years old.

Physical features

The fossil record of *Homo erectus* from East and Southeast Asia is dominated by skulls, jaws, and teeth. The few remains from other parts of the body are fragmented, show signs of disease, or are insecurely dated. Taking this into account, it appears that *Homo erectus* was large-bodied, with fully modern bipedalism and a relatively large brain. The face and cranial vault were distinctive, with a strong brow region and wide cheekbones. Morphological variation in this species may be attributable to differences between the sexes, regional changes, or variation through time.

> **HEIGHT** 5 ft 3 in–6 ft (1.6–1.8 m)

> **WEIGHT** 88–150 lb (40–68 kg)

> **BRAIN SIZE** 46–79 cubic in (750–1,300 cubic cm)

occipital bone more angled than in modern humans

long, low cranial vault

eye sockets small and rectangular

skull widest toward the base

SANGIRAN SKULL
Discovered in 1969, this specimen (Sangiran 17), is the most complete *Homo erectus* skull yet found in Java. The cranial vault is long and low with a relatively large brain capacity of around 61 cu in (1,000 cc). The face is wide, with forwardly placed cheekbones and a thick, straight brow ridge.

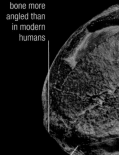

pathological bony growth probably caused by injury

LOWER BODY Few fossil bones of the lower body are known, but the small collection that does exist includes a complete thigh bone (Trinil 3). The size and shape of Trinil 3 is distinctly modern, with a tear-shaped shaft cross section, a muscle attachment site running down the back, and large joint surfaces.

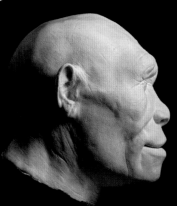

SKULL The many partial *Homo erectus* skulls and teeth discovered are quite variable through time and space. Overall, skulls are characterized by a large but long and low cranial vault topped with a sagittal crest, and by a strongly angled occipital bone to the rear. Cranial capacity varies but averages 61 cu in (1,000 cc). The face is large and wide, with pronounced cheekbones. However, it is quite vertical, the nose is prominent, and jaws and teeth are relatively small. Compared to a modern human's, the face is dominated by the low forehead and a massive browridge that may be either straight or arched over the eye sockets.

"PEKING MAN"
Many *Homo erectus* remains come from China, including about 40 from Zhoukoudian, and were once considered a separate species (*Homo pekinensis*). Many original fossils were lost during the second Sino-Japanese war, with only casts and descriptions surviving.

thick keel of bone runs along midline of skull

skull narrows behind eye sockets

wide cheekbones

cheek teeth larger than in modern humans

receding chin

Archaeology

The stone tool kits that accompany the fossils of *Homo erectus* in Asia are strikingly lacking in the large handaxes that are so common from around 1.65 million years ago in Africa, and later also in Europe—although the earliest hominins in Europe did not use handaxes either. Did hominins leave Africa before handaxes were invented, or did they or forget how to make them? Or were other kinds of tools made from perishable materials, such as bamboo or shell, more appropriate to the local environments? In 2015, analysis of an old collection of shells from the type locality of *Homo erectus* at Trinil, Java, showed that at least one appeared to have been used as a tool. More remarkably, another shell bore a geometric series of scratch-marks apparently made deliberately, without any obvious functional purpose.

These marks may represent the earliest deliberate production of aesthetic or even symbolic artifacts by a hominin, suggesting *Homo erectus* was capable of highly complex thoughts and behaviors.

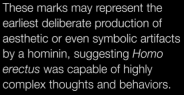

CHINESE TOOLS
Excavations at Bose have revealed the oldest large cutting tools in China. These required similar production skills to handaxes.

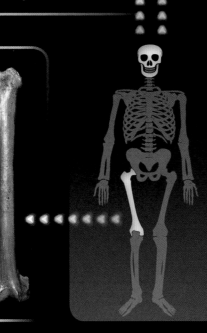

large, low skull

strong, shelflike
brow ridge

Reconstruction

This adult male *Homo erectus* has been reconstructed from the famous skull
known as Sangiran 17. The skull was found on the Indonesian island of Java
in 1969, and it dates back to more than 1 million years ago. A large part
of the face of the skull is badly damaged, so this reconstruction uses parts of
the facial skeleton and upper jaw of another specimen (Sangiran 4) from the
same site; the lower jaw is modeled on a 700,000-year-old specimen found
in Algeria, at the site of Tighenif (also known as Ternifine). The modelers have
given this individual a similar skin tone to that of modern-day Indonesians.

FRONT RIGHT

FRONT

smaller irises and distinctive whites to the eyes may have evolved by this time, aiding communication by eye movement

dark skin was probably a characteristic of the genus *Homo*, evolving as an adaptation to our ancestors' loss of body hair and the exposure of their skin to the sun

nasal bones—and hence the nose—are broad, and the bridge of the nose is quite flat

jaw still protrudes farther out from under the face than in modern humans, and is heavy and solidly built

FRONT LEFT

SIDE LEFT

strong chewing muscles running up the sides of the head were anchored by a bony "keel" along the top of the skull

wide, flat cheekbones contribute to a broad face

Homo
antecessor

Although *Homo antecessor* is a controversial species, its fossils revealed that hominins reached Western Europe at least 780,000 years ago.

Homo antecessor may be the first western European hominin species. It is controversial, because the description is based on an immature specimen, and many of the features used to define it are shared with other species, especially *Homo heidelbergensis*.

Discovery

The Sierra de Atapuerca hills, near Burgos in northern Spain, contain more than 2.5 miles (4 km) of limestone caves. Many of these caves are filled with fossil-rich sediment. During the late 19th century, a long trench was cut through the southwestern part of the hills to make a cutting for a railroad, exposing sediment-filled fissures. The archaeological importance of this trench (known as the Trinchera del Ferrocarril) was not realized until the 1960s, when a team including staff from the Museum of Burgos and members of the Edelweiss Speleological Group began a survey of the site and discovered stone tools. In 1972, the graduate student Trinidad Torres excavated one northern area, which he named Gran Dolina. From 1978 to the present, Gran Dolina has been the focus of meticulous archaeological excavation, directed first by Emiliano Aguirre and more recently Juan Luis Arsuaga, José Maria Bermudez de Castro, and Eudald Carbonell. In 1994, the first hominin finds were made. They consisted of immature jawbones and 11 teeth from

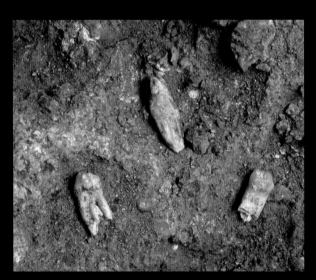

Gran Dolina's oldest layer, known as TD-6, which is securely dated to at least 780,000 years ago. These findings were published in 1997 as the type specimen for a new species, named *Homo antecessor*. Since then, many more fragmented fossils have been discovered, and the *Homo antecessor* population is now represented by 8–14 individuals.

ANTECESSOR TEETH
These ancient hominin teeth, shown in situ, were among 11 excavated from level TD-6 of the Gran Dolina archaeological site in 1994. Three years later, they were attributed to *H. antecessor*. The teeth of this species are more robust than those of modern humans.

HOMO ANTECESSOR

> **NAME MEANING** "Pioneer Man"

> **LOCALITY** Atapuerca, northern Spain

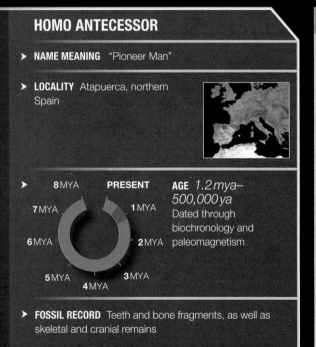

8 MYA	PRESENT	1 MYA
7 MYA		
6 MYA		2 MYA
5 MYA		3 MYA
	4 MYA	

AGE *1.2 mya–500,000 ya*
Dated through biochronology and paleomagnetism

> **FOSSIL RECORD** Teeth and bone fragments, as well as skeletal and cranial remains

POSSIBLE CANNIBALISM

The *Homo antecessor* fossils from TD-6 were found disarticulated and mixed with the bones of other species. Most of these animal and hominin remains show marks from cutting, chopping, and striking with stone tools. Archaeologists have speculated that hominin corpses were butchered for consumption, and that the bones were discarded in just the same way as those of hunted or scavenged animals.

CUT MARKS
This hominin hand bone (ATD6-59) shows cut marks on the shaft at a muscle attachment site.

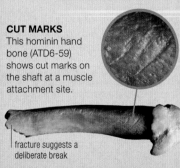

fracture suggests a deliberate break

SKULL There are no complete *Homo antecessor* skulls, but partial jaws, faces, braincases, and teeth have been discovered. The most complete specimen (ATD 6-69) includes a child's upper jaw and cheekbone, but there are also several adult fragments in the collection. Overall, the most striking feature of the child's skull is the fully modern mid-face, with its high-set cheeks. Above the fairly widely spaced eye sockets, the brow ridge is modest and arched, and the maxillary sinuses (in the cheekbones) are large. There is a marked nasal prominence, and the upper part of the nasal opening is narrow.

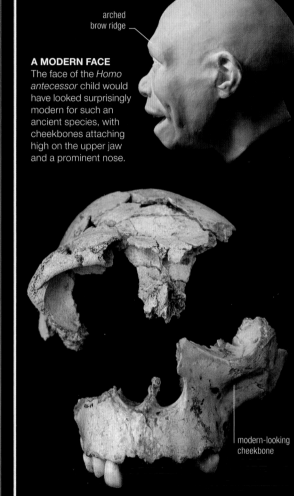

arched brow ridge

A MODERN FACE
The face of the *Homo antecessor* child would have looked surprisingly modern for such an ancient species, with cheekbones attaching high on the upper jaw and a prominent nose.

modern-looking cheekbone

CHILD'S JAW AND TEETH
This upper jaw has some adult teeth, a canine and premolar just erupting, and back molars still hidden in their sockets. The child was probably 10–11.5 years old when it died.

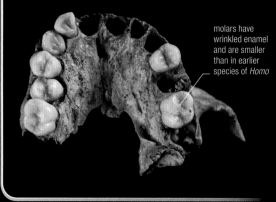

molars have wrinkled enamel and are smaller than in earlier species of *Homo*

Physical features

No complete skeletons are known for *Homo antecessor*, but some regions of the body—including the skull, thorax, and feet—are quite well represented in the disarticulated (jumbled) fossil collection from Gran Dolina. Together, these fossils suggest that the *Homo antecessor* population was of a similar average stature to modern humans and fully bipedal, but with longer, more slender arms and wider chests. The skull of this species was rounded and quite lightly built, with an average brain capacity of approximately 61 cu in (1,000 cc). *Homo antecessor*'s face would have had a remarkably modern appearance, including a prominent nose, but receding chin.

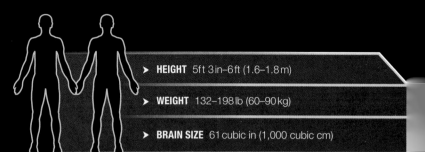

> **HEIGHT** 5ft 3in–6ft (1.6–1.8m)

> **WEIGHT** 132–198lb (60–90kg)

> **BRAIN SIZE** 61cubic in (1,000 cubic cm)

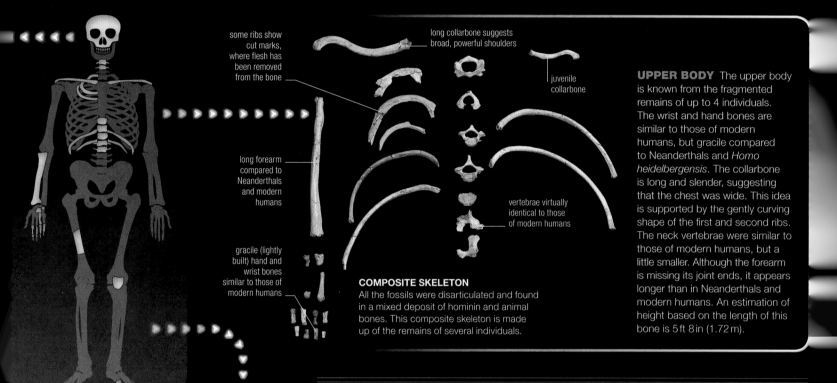

some ribs show cut marks, where flesh has been removed from the bone

long collarbone suggests broad, powerful shoulders

juvenile collarbone

long forearm compared to Neanderthals and modern humans

vertebrae virtually identical to those of modern humans

gracile (lightly built) hand and wrist bones similar to those of modern humans

COMPOSITE SKELETON
All the fossils were disarticulated and found in a mixed deposit of hominin and animal bones. This composite skeleton is made up of the remains of several individuals.

UPPER BODY The upper body is known from the fragmented remains of up to 4 individuals. The wrist and hand bones are similar to those of modern humans, but gracile compared to Neanderthals and *Homo heidelbergensis*. The collarbone is long and slender, suggesting that the chest was wide. This idea is supported by the gently curving shape of the first and second ribs. The neck vertebrae were similar to those of modern humans, but a little smaller. Although the forearm is missing its joint ends, it appears longer than in Neanderthals and modern humans. An estimation of height based on the length of this bone is 5ft 8in (1.72m).

LOWER BODY Overall, it is argued that the *Homo antecessor* skeleton was more similar to modern humans than to other Middle and Upper Pleistocene European hominins. The lower body is represented by the upper part of a thigh bone shaft, two kneecaps, and various foot bones. Unlike the rest of the lower limb, the thigh bone resembles those of Neanderthals and earlier hominins more than that of *Homo sapiens*.

kneecaps are relatively narrow, more like modern humans' than Neanderthals'

MODERN FEET
The feet of *Homo antecessor* were similar to those of modern humans, although the bone of the big toe may have been more rounded in shape.

narrower mid-foot bones (metatarsals) than Neanderthals

toe bones (phalanges) closely resemble those of modern humans

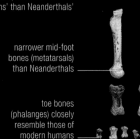

Archaeology

Stone tools have been found associated with *Homo antecessor* at Gran Dolina. Excavations at Happisburgh and Pakefield in the UK have also produced stone tools dated to a similar time period. Although no hominin remains have been found—though fossil footprints were found at Happisburgh —it is thought that these were probably also made by *H. antecessor*. Like the African Oldowan, these tools include hammerstones, small flakes, and cores, and lack the handaxes of later stone tool technologies. The presence of tiny stone fragments produced by knapping demonstrates that some of the artifacts were made on site from local materials, possibly for cutting meat and woodworking.

LOCAL RAW MATERIALS
Gran Dolina, about 9 miles (15km) from the city wof Burgos, is one of Europe's oldest hominin sites. All the stone used to make tools at Gran Dolina can be sourced within a 3km (1.9mile) radius of the site.

CORES AND FLAKES
Hominins brought some large blocks of stone to Gran Dolina to produce tools on site. By refitting stone fragments together like a jigsaw puzzle, archaeologists can reconstruct the tool production sequence.

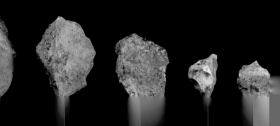

hair conceals
protruding bony
"bun" at back
of skull

more modern,
flatter shape to
midfacial region
than in earlier
hominins

jaw probably
protruded farther than
in modern humans

Reconstruction

As yet, no complete *Homo antecessor* skull has been found, so this
reconstruction is based on two fossils from different individuals, both around
800,000 years old. One, ATD 6-15, preserves the front part of the braincase
and a corner of one eye socket; the other, ATD 6-69, comes from a young
individual, perhaps around 10 years old, and includes the upper jaw and the
lower part of one cheekbone and bottom of the other eye socket. The jaw is
based on a Neanderthal specimen—although the Neanderthals lived long
after *H. antecessor*, their mandibles share many similarities.

FRONT RIGHT

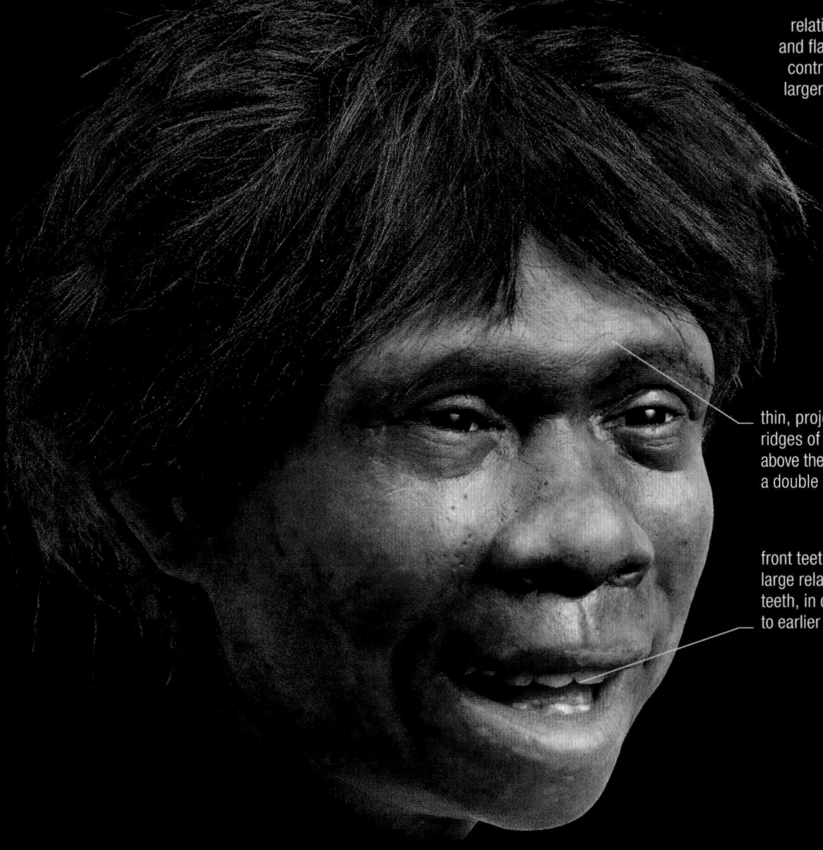

FRONT

relatively broad and flat forehead contributes to a larger braincase

thin, projecting ridges of bone above the eye form a double arch

front teeth are quite large relative to back teeth, in comparison to earlier hominins

upper front teeth have a distinctive "shovel" shape, similar to that of some *Homo erectus* and modern human populations

FRONT LEFT

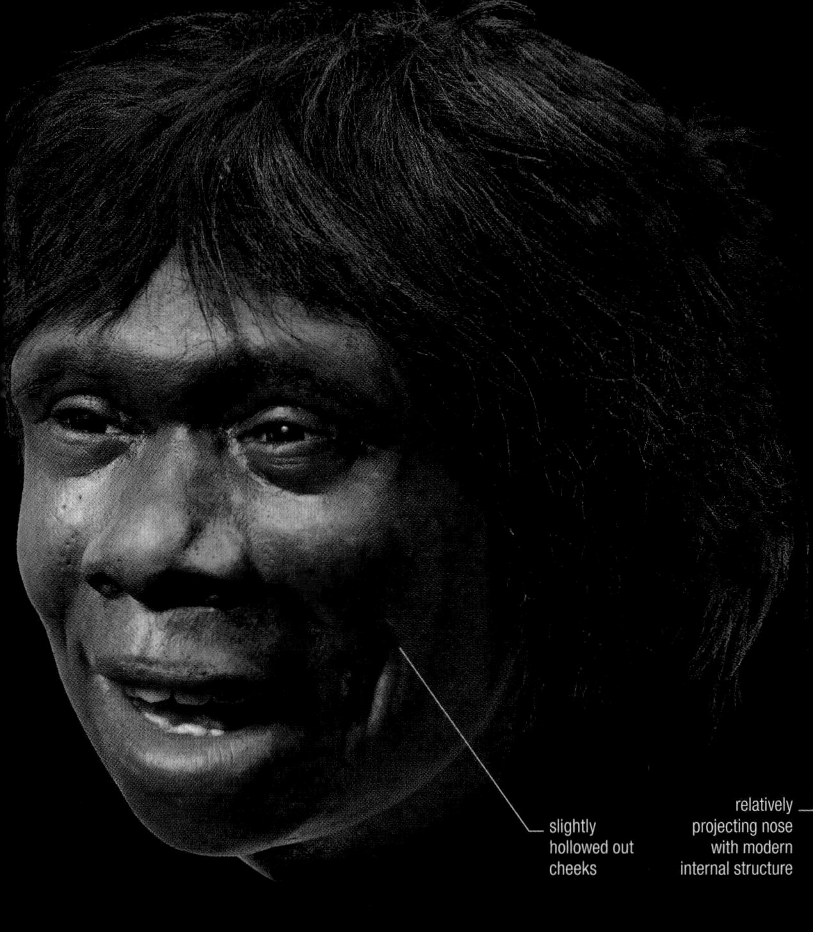

SIDE LEFT

slightly hollowed out cheeks

relatively projecting nose with modern internal structure

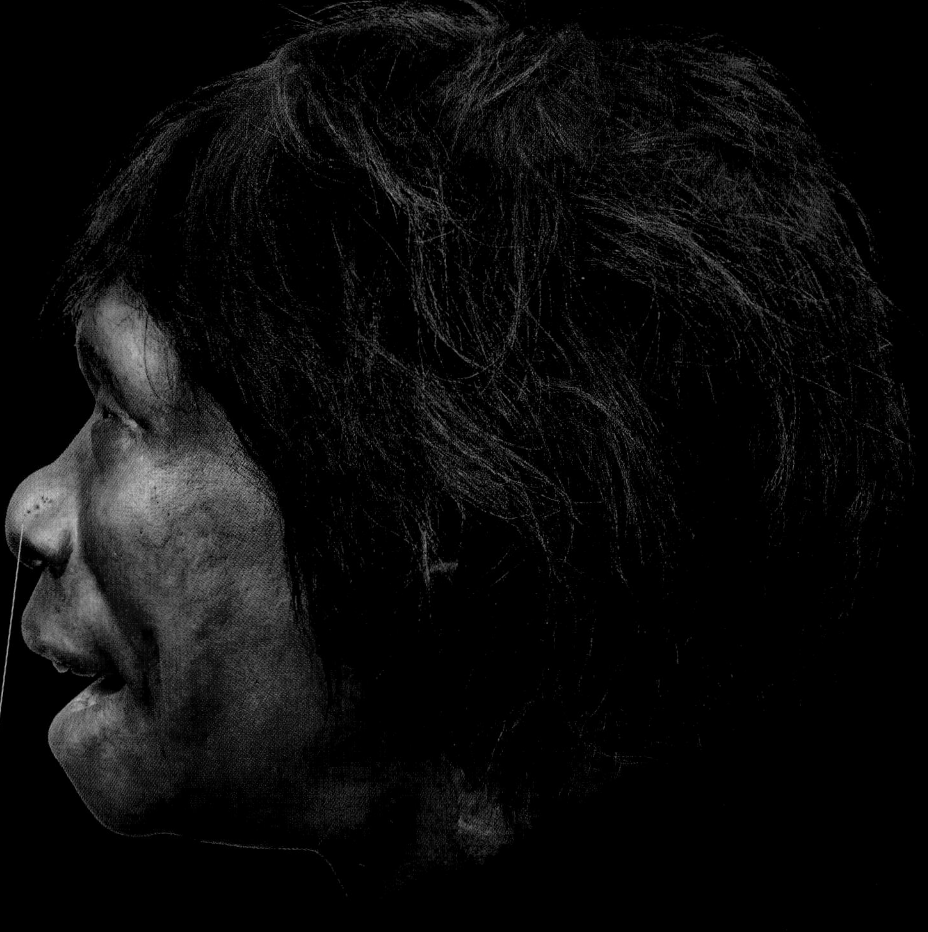

Homo
heidelbergensis

This species was probably the last common ancestor of Neanderthals, in Europe, and modern humans, in Africa.

Homo heidelbergensis seems to have lived over a wide area, from southern Africa to northern Europe. These early people had large brains, strong muscular bodies, and the ability to hunt large animals and create relatively complex toolkits.

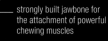

strongly built jawbone for the attachment of powerful chewing muscles

smaller molars than in early hominins

Discovery

In October 1907, a workman excavating a sand quarry near the village of Mauer, southeast of Heidelberg in Germany, unearthed a robust human-like lower jaw. The fossils of many extinct species had already been found in the same 79ft- (24m-) deep deposits. In 1908, the jaw was published by the excavation director Otto Schoentensack of the University of Heidelberg. He assigned the specimen a new species name, *Homo heidelbergensis*, and almost immediately it became the subject of controversy, as scientists began to explore its relationship to *Homo neanderthalensis*. Gradually, more Middle Pleistocene fossils were discovered and drawn into the debate. By the 1970s, they included mandibles and a cranium from Arago (France), a cranium from Petralona (Greece), and a cranium from Kabwé in Zambia (originally described as *Homo rhodesiensis*). The fossils show both *Homo erectus*-like features and modern human characteristics.

MAUER MANDIBLE
The type specimen for this species is a well-preserved lower jaw, or mandible, from Mauer. It is highly robust and has a receding chin, but relatively small molars similar to those of modern humans.

ARAGO CAVES
Since 1964, excavations at Arago near Tautavel, southwestern France, have revealed a number of *Homo heidelbergensis* fossils, including the near-complete skull of a young male.

A controversial species

Originally applied only to the European fossils, the term *Homo heidelbergensis* was later used to describe all fossils from both Europe and Africa from between 600,000 and 300,000 years ago, making this species the common ancestor of Neanderthals (in Europe) and *Homo sapiens* (in Africa). However, despite their similarities, there are also clear regional differences. Many anthropologists believe the remains belong to more than one species, probably descended from *Homo ergaster* or *Homo erectus*, and that the African finds should not be included in *H. heidelbergensis* but ascribed to a different species called *Homo rhodesiensis*, which ultimately evolved into early *H. sapiens* in Africa 300,000–200,000 years ago. In contrast, European *H. heidelbergensis* evolved into the Neanderthals—and possibly also the Denisovans—more than 400,000 years ago. Fossils from the Sima de los Huesos site in Spain, previously thought to have been *Homo heidelbergensis*, have been shown to be early Neanderthals on the basis of both bone and DNA analysis. Nevertheless, these remains of around 30 individuals provide the most complete picture of *Homo* anatomy before the emergence of definitive Neanderthals and modern humans.

large nasal opening and upper jaw project forward

large, heavily built pelvis

femur long and straight, with a round cross-section

lower leg bones have thick shafts, indicating high levels of activity

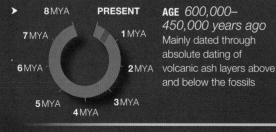

JUMBLED SKELETONS
The bones of many of the Sima de los Huesos individuals were found jumbled up at the base of the cave shaft, as if they were thrown or washed in. Some skeletons have now been reassembled.

Physical features

Homo heidelbergensis is known from a number of skulls, jaws, and other fragments found across Europe and Africa. Few body and limb remains have been found, making it difficult to establish average heights and sizes in this species. However, the Boxgrove tibia (see below) belonged to an individual around 6ft (1.93m) tall, and the size and heavy build of the fossil skulls suggests that individuals were robust and relatively tall. Such strongly built individuals would have been well-suited to hunting large herd animals in frequently harsh Pleistocene environments.

> **HEIGHT** 4ft 9in–5ft 11in (1.45–1.80m)

> **WEIGHT** 110–196lb (50–89kg)

> **BRAIN SIZE** 67–85 cubic in (1,100–1,400 cubic cm)

SKULL Large and long, with strong attachment sites for neck muscles, the skull accommodated a big brain of around 78 cu in (1,274cc)—slightly smaller than in Neanderthals and within the range of modern humans. The broad, relatively flat face has strong brow ridges running above the wide-set eyes. The large jaws provide plenty of space for the modestly sized teeth, which are slightly larger in males of the species.

large brow ridges form distinct arches over each eye

long, sloping forehead

KABWÉ SKULL
This skull from Zambia is one of the most robust assigned to Homo heidelbergensis, with a massive brow ridge and thickened bone at the back to support the strong neck muscles.

LOWER BODY The tibia shaft (shin bone) of an adult male found at Boxgrove, UK, in 1993 is one of the few well-preserved limb fossils known attributed to Homo heidelbergensis. The bone wall that encloses the inner marrow is very thick compared to that of modern humans, suggesting it was under a great deal of biomechanical stress, probably caused by walking long distances while hunting.

GNAWED BONE
The shin bone was gnawed by a carnivore, probably a wolf. Whether the animal killed this individual or it scavenged the bone is not certain.

Archaeology

Homo heidelbergensis is largely associated with stone tools manufactured using the prepared-core technique. This technique, shared with Neanderthals and early Homo sapiens, allowed flakes of a predetermined size and shape to be struck off a preshaped core with a single blow. Tools were often made of fine-grained rock with good flaking properties, and toolkits show a variety of forms, including handaxes, points, and flakes. Some of these were used for hunting and butchery, and are found with animal bones.

Surviving in the north

Homo heidelbergensis colonized Europe in the Middle Pleistocene. As these early people moved into Europe, they faced new challenges. Not only were northern landscapes colder, but there were also marked differences in the climates and resources available in different seasons. However, the open grasslands of Pleistocene Europe were home to large herds of mammals, which presented new resources to exploit.
H. heidelbergensis became an expert hunter, using not only finely made handaxes but also innovative tool forms, such as the wooden thrusting spears found at Schoningen and Lehringen in Germany and Clacton in the UK (see p.182). These new hunting technologies allowed the species to compete successfully with large carnivores, whose tooth marks also appear on the bones of animals killed by H. heidelbergensis at Boxgrove, indicating the animals scavenged carcasses that people had hunted.

BOXGROVE

About 500,000 years ago, Boxgrove in West Sussex, UK, was the site of a tidal lagoon, a beach, and limestone cliffs. Excavations in a local quarry led by archaeologist Mark Roberts revealed a rich record of early hominin occupation. Acheulean artifacts, including very finely knapped handaxes, were found in association with mammals such as horses, giant deer, and rhinoceroses, which may have been hunted while they visited the lagoon. Two lower incisors and a partial shin bone provide evidence of the hominin colonizers themselves.

RHINO TOOTH
One of the fossils found at Boxgrove was the tooth of an extinct rhinoceros, Stephanorhinus, which was about the same size as the modern-day white rhino.

low, sloping
forehead

pigments
may have
been used
for body
decoration

wide and
relatively
pronounced
nose

Reconstruction

This adult male *Homo heidelbergensis* is based on the skull found in 1921 at
Broken Hill (Kabwé) in Zambia. Originally dated to 40,000 years old, the skull was
originally assigned to *Homo rhodesiensis*, although others believed it to be an
African Neanderthal or a larger type of *Homo erectus*. It was used to argue that
evolution in Africa lagged behind that in Europe. The skull has since been
re-dated to at least 125,000 years old, perhaps as much as 300,000. The lower
jaw is based on one found at Mauer, near Heidelberg in Germany—the
"type-fossil," the standard to which other fossils of this species are compared.

FRONT RIGHT

FRONT

sides of the skull slope together into a pinched "keel" to the top and back

Homo heidelbergensis lived in both Africa and in more northerly latitudes in Europe, but probably retained the dark skin of their African ancestors

this individual suffered from severe dental decay; an infected abscess in his upper jaw may have led to his death

FRONT LEFT

SIDE LEFT

substantial brow ridges

partially healed skull wound was found just above the left ear; it may have been caused by a sharp implement or a carnivore

teeth were smaller than those of earlier species

robust lower jaw with receding chin

Homo *floresiensis*

Although some scientists believe that this tiny hominin was a modern human with a physical disorder, recent evidence suggests that *Homo floresiensis* was a distinct species.

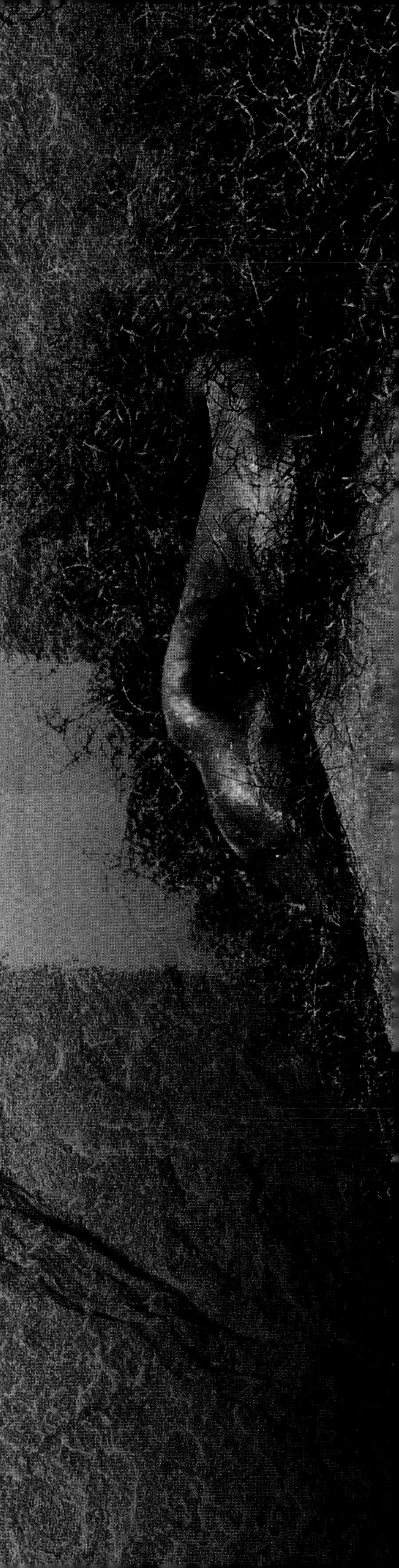

Homo floresiensis was a very small hominin that lived between 700,000 and 60,000 years ago on the Indonesian island of Flores. It remains unclear how and why the species became so small or how it fits into the hominin family tree.

Discovery

Liang Bua, or the "cool cave", was first identified as an archaeological site in the 1960s, and a series of excavations during the 1970s and 1980s revealed that the cave was used by modern humans over the past 10,000 years. In 2001, joint Indonesian-Australian excavations started to investigate whether there had been any earlier hominin occupation of the site. In 2003, the team made the remarkable find of the partial skeleton of a tiny female hominin (LB1) buried underneath a thick volcanic-ash layer. LB1 was initially dated to just 18,000 years ago—very recent in terms of human evolution. LB1's unique combination of features, including a very short stature and small brain size led to the remains being assigned to a new species, _Homo floresiensis_, in 2004. This was controversial, as many anthropologists were concerned that finding such traits in such a recent specimen suggested that the LB1 individual might have been a modern human, but one that was diseased or suffered from a hormonal disorder. However, since the initial discovery, other fragmented specimens with similar traits have been found—including remains from the nearby site of Mata Menge—and the Liang Bua remains have been re-dated to 100,000–60,000 years ago. Although the finds remain controversial, there is now more general agreement that this ancient population was distinctive enough to merit being classified as a new species.

ISLAND ISOLATION
The island of Flores has long been separated from Australia and Asia. Its isolation could, over a long period of time, have lead to hominin species living here "shrinking" in size when compared to their mainland relatives—a phenomenon known as island (or insular) dwarfism.

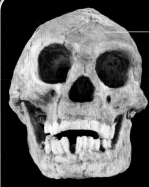
brain size was like that of a living chimpanzee

long, low cranium, closer to that of _H. erectus_ than _H. sapiens_

SMALL BRAIN
The brain size was similar to the that of smallest australopithecine brains. However, the brain had a flattened shape, top to bottom, and an anatomy that may be more similar to _Homo_.

small, receding forehead

SMALL FACE
The face is reduced and set quite vertically under the frontal bone, with rounded eye sockets and arched brow ridges.

SKULL _Homo floresiensis_ has a very small skull for a _Homo_ species. The thick-walled braincase is globular, with a low profile. The nasal opening is close to the narrow, slightly protruding upper tooth row, and the canines are small. The chin recedes, but the lower jaw is strong, with distinct muscle attachment markings.

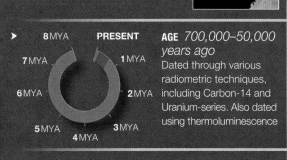

HOMO FLORESIENSIS

> **NAME MEANING** "Flores Man"

> **LOCALITY** Liang Bua Cave on the island of Flores, Indonesia

> **AGE** _700,000–50,000 years ago_
> Dated through various radiometric techniques, including Carbon-14 and Uranium-series. Also dated using thermoluminescence

8 MYA · PRESENT · 1 MYA · 7 MYA · 2 MYA · 6 MYA · 3 MYA · 5 MYA · 4 MYA

> **FOSSIL RECORD** One nearly complete skull and partial skeleton, and parts of at least 11 other individuals

WHAT IS HOMO FLORESIENSIS?

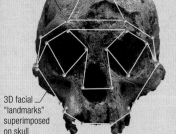

skull quite narrow

3D facial "landmarks" superimposed on skull

In 2009, a digital study using 3-D shape analysis compared different regions of the LB1 skull to those of other extinct hominins, and also _Homo sapiens_ and apes. The results indicated that it showed more similarities to _Homo_ species that lived around 1.5 million years ago than to modern humans.

LOWER BODY The lower limbs are relatively short and stocky, with wide shafts in cross section. Muscle attachment sites on the pelvis and thigh bone (femur) indicate that the legs were oriented differently from those of modern humans, and the joint surface of the hip is small. The blade of the pelvis flares out to the side and is wider than that of an average modern human. Research suggests that this species may have been a poor runner, perhaps using the arms and upper body in locomotion more than modern humans do. Whether these characteristics were caused by disease, island dwarfism, or simply the normal anatomy of a small human population is still uncertain.

LONG FEET
LB1's feet show a combination of primitive and modern traits, suggesting that its gait was different from our own. The big toe is set parallel to the other toes and the bones of the mid-foot have a similar pattern of robusticity to those of modern humans, but the arches are less developed and the foot is very long in comparison to the length of the leg.

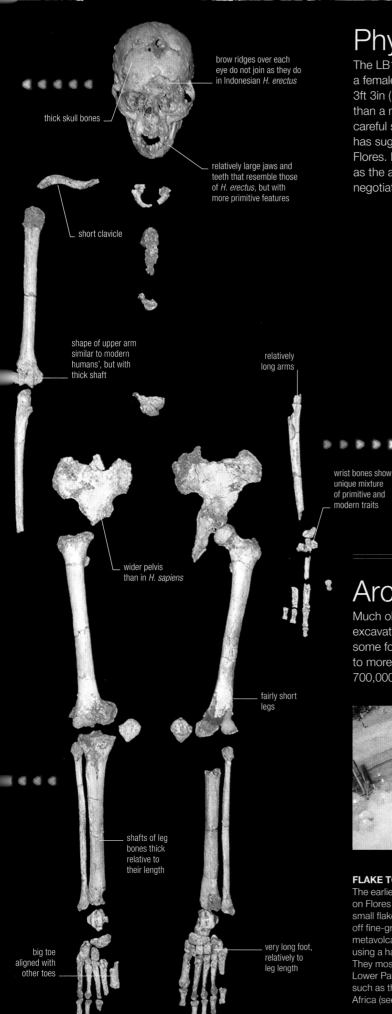

thick skull bones

brow ridges over each eye do not join as they do in Indonesian *H. erectus*

relatively large jaws and teeth that resemble those of *H. erectus*, but with more primitive features

short clavicle

shape of upper arm similar to modern humans', but with thick shaft

relatively long arms

wrist bones show unique mixture of primitive and modern traits

wider pelvis than in *H. sapiens*

fairly short legs

shafts of leg bones thick relative to their length

big toe aligned with other toes

very long foot, relatively to leg length

Physical features

The LB1 skeleton, the most complete *Homo floresiensis* fossil so far discovered, is that of a female whose teeth suggest that she died in early middle age. Standing only a little over 3ft 3in (1m) tall and with a brain size of about 24 cubic in (400 cubic cm), she was smaller than a modern chimpanzee. No DNA is preserved in the remains from Liang Bua, but careful study of the physical similarities and differences between LB1 and other hominins has suggested that it may descend from a group of *Homo erectus* that became isolated on Flores. However, other work has pointed to *Homo habilis* or even *Australopithecus sediba* as the ancestor of LB1, implying a much earlier dispersal out of Africa—and successful negotiation of the sea-crossing to Flores—than was previously thought possible.

> **HEIGHT** Female: 3 ft 7 in (1.1 m)

> **WEIGHT** 55 lb (25 kg)

> **BRAIN SIZE** 23–26 cubic in (380–420 cubic cm)

UPPER BODY Although short in stature, *Homo floresiensis* had robust limbs and probably a strong, muscular body. The LB1 skeleton possesses relatively long arms but short legs, and the relative proportions of the limbs are closer to those of *Australopithecus* species and *H. habilis* than modern humans. The bone of the upper arm (humerus) has a modern shape, but the collarbone (clavicle) is short and the shoulder joint is arranged more like that of *H. erectus* than *H. sapiens*. The cross section of the humerus shaft is wide, which may suggest considerable upper body strength, making it more similar to early humans and apes than to modern humans.

Archaeology

Much older evidence of hominin occupation of Flores than that from Liang Bua comes from excavations at Wolo Sege, Boa Lesa, and Mata Menge in the Soa Basin. Here stone tools, some found next to the bones of animals such as the elephant Stegodon, have been dated to more than 1 million years years old. At Mata Menge fragments of jaw and teeth dated to 700,000 years ago resemble those found at Liang Bua and may represent LB1's ancestors.

There was never a land bridge connecting Flores to the mainland, so hominins must have made a dangerous sea crossing to reach the island.

EARLY OCCUPATION OF FLORES
Joint Indonesian-Australian-European excavations have discovered hominin sites dating back to 1 million years ago, but which species created them remains controversial.

FLAKE TOOLS
The earliest stone tools on Flores are mostly small flakes knapped off fine-grained metavolcanic cobbles using a hard hammer. They most resemble Lower Paleolithic tools such as the Oldowan of Africa (see p.102).

LIANG BUA CAVE
This 130 ft- (40 m-) wide limestone cave, situated at the western end of the tiny Indonesian island of Flores, is thought to have formed up to 400,000 years ago. It is the only site where *Homo floresiensis* fossils have been found.

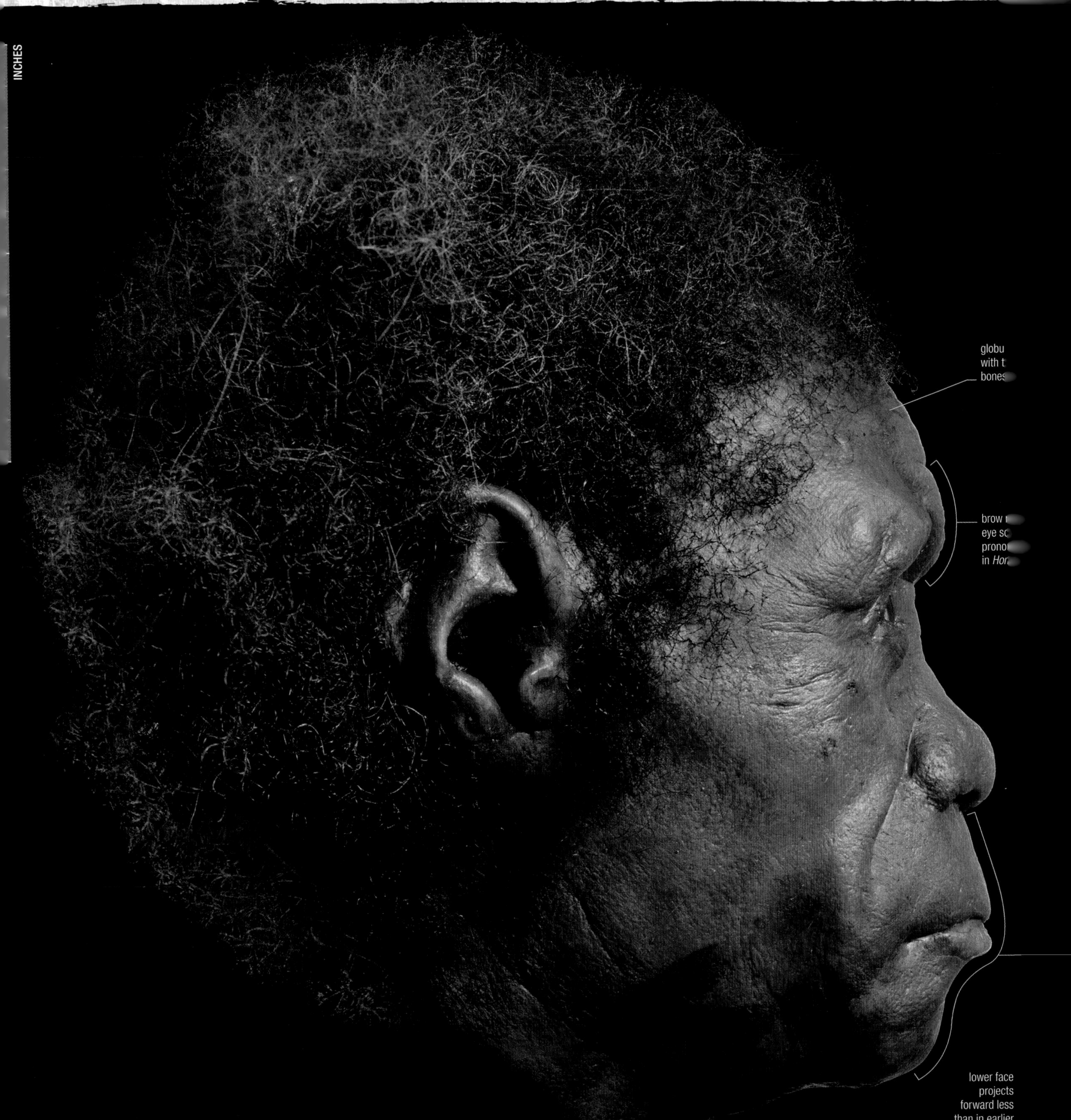

INCHES

globu
with t
bones

brow
eye sc
prono
in *Hor*

lower face
projects
forward less
than in earlier
species

Reconstruction

This reconstruction of *Homo floresiensis* is based on the relatively complete
skeleton and skull known as LB1 found in the Liang Bua Cave (see pp.142–45),
between 2003 and 2004. Close study of the shape of the skull and the imprints
of the brain left on its inside suggest it is a genuine early hominin rather than
a recent modern human who suffered from a genetic condition or disease
that restricted its growth. The remains are associated with archaeology that
suggests relatively complex behaviors, calling into question many of our
assumptions about the relationships between brain size and intelligence

FRONT RIGHT

minute brain-case compared to other species of *Homo*

nose and mouth separated by deep upper lip

jaw shape and size less like that of *H. erectus* or *H. sapiens* and much more similar to *H. habilis* or *H. rudolfensis*

FRONT

cheekbones less convex and pronounced than in australopithecines and paranthropines

LB1's face was slightly lopsided—possibly a sign of disease, developmental disorder, or maybe a degree of lopsidedness within normal range for primates

FRONT LEFT

eyes set relatively widely apart

smaller and narrower face than earlier hominin species

dark pigmentation probably helped protect skin from midlatitude sun

SIDE LEFT

receding chin, as in earlier hominins

Homo
neanderthalensis

The Neanderthals thrived in Europe for around 300,000 years before modern humans arrived. The reason for the demise of this successful species remains a mystery.

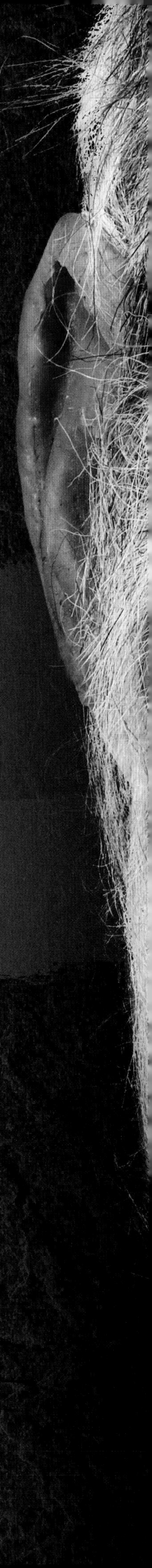

Homo neanderthalensis was the first fossil hominin discovered and described. Today, we have thousands of fossil specimens excavated from hundreds of sites across Europe. This record is remarkable, because it includes individuals of all ages, from premature fetuses to the very elderly.

Discovery

The first two Neanderthal finds—infant remains found near Engis, Belgium, by Charles Schmerling in 1829, and a skull discovered at Forbes Quarry, Gibraltar, in 1848—were not recognized as such for many years. This changed in 1856, after workers at the Feldhofer cave in the Neander Valley, Germany, discovered a partial human skeleton and passed it to local naturalist Johann Fuhlrott. The skull's unusual features convinced him that he had found a fossil human, and in collaboration with the anatomist Hermann Schaaffhausen the find was published in 1857. Seven years later, geologist William King suggested the name _Homo neanderthalensis_, and zoologist George Busk made the link between the Gibraltar and Feldhofer specimens. From then on, discoveries of _H. neanderthalensis_ were made regularly, and in 1936 paleontologist Charles Fraipont finally correctly identified the Engis infant as a Neanderthal.

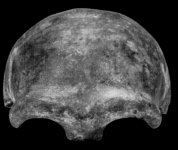

NEANDERTHAL 1
This skull is from a partial skeleton that was the first specimen to be identified as a human fossil. The remains were described in 1857—two years before Darwin's _On the Origin of Species_.

NEANDER VALLEY
This valley, near Düsseldorf in Germany, was a limestone canyon, but in the 19th century industrial mining removed much of the rock. Although the limestone-cliff cave site where Neanderthal 1 was found in 1856 was destroyed, subsequent excavations during the 1990s on the ground where the cliff once stood have produced more fossil remains.

The remains of more than 275 individual Neanderthals have been discovered, from more than 70 sites across Europe and western Asia. The Neanderthal skeleton known as La Ferrassie 1, found in France in 1909, is the basis for much of this reconstruction. The pelvis, rib cage, and backbone were modelled from other skeletons.

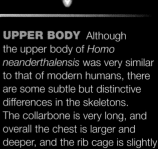

UPPER BODY Although the upper body of _Homo neanderthalensis_ was very similar to that of modern humans, there are some subtle but distinctive differences in the skeletons. The collarbone is very long, and overall the chest is larger and deeper, and the rib cage is slightly flared at the base. The strong arms show marked muscle attachment sites. The length ratio of forearm to upper arm is lower in Neanderthals than in modern humans, reflecting the relatively short Neanderthal forearm. The fingers are short, but the fingertips are broad. The scapula (shoulder blade) is wide from side to side, and it is likely that Neanderthals had a powerful upper arm swing.

HOMO NEANDERTHALENSIS

> **NAME MEANING** "Human from the Neander Valley"

> **LOCALITY** Right across Europe and into Siberia, and also into southwest Asia

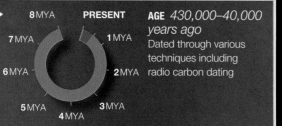

> 8 MYA PRESENT **AGE** _430,000–40,000 years ago_
> 7 MYA 1 MYA Dated through various
> 6 MYA 2 MYA techniques including radio carbon dating
> 5 MYA 3 MYA
> 4 MYA

> **FOSSIL RECORD** A number of entire skeletons, plus various bone fragments from more than 275 individuals

NEANDERTHAL CHILDREN

The first Neanderthal specimen discovered was that of an infant just under 2 years old. Subsequently, the remains of more than 100 fetuses, babies, children, and adolescents have been excavated. Such high numbers of juveniles are unique among nonhuman hominins, and some suggest that this represents a high level of infant mortality. Although some studies have indicated that Neanderthal children may have matured relatively quickly, recent analysis of a partial juvenile skeleton from El Sidron in Spain suggests a similar growth rate to that of modern human children.

ROC DE MARSAL
In 1961, the partial skeleton of a Neanderthal child about 3 years old was discovered at Roc de Marsal, near Les Eyzies, France.

LOWER BODY Neanderthals were a little shorter in the lower body than modern humans. The pelvis is wide; at the front, the two pubic bones are very long, and the joint surface between them is tall but narrow. The waist area is short, because there is limited space between the top of the pelvis and the thorax. The thigh bones and shin bones are robust, with large muscle attachment areas and slightly bowed shafts. The length ratio of lower leg to thigh bone is lower than in modern humans, reflecting the relatively short shin bone.

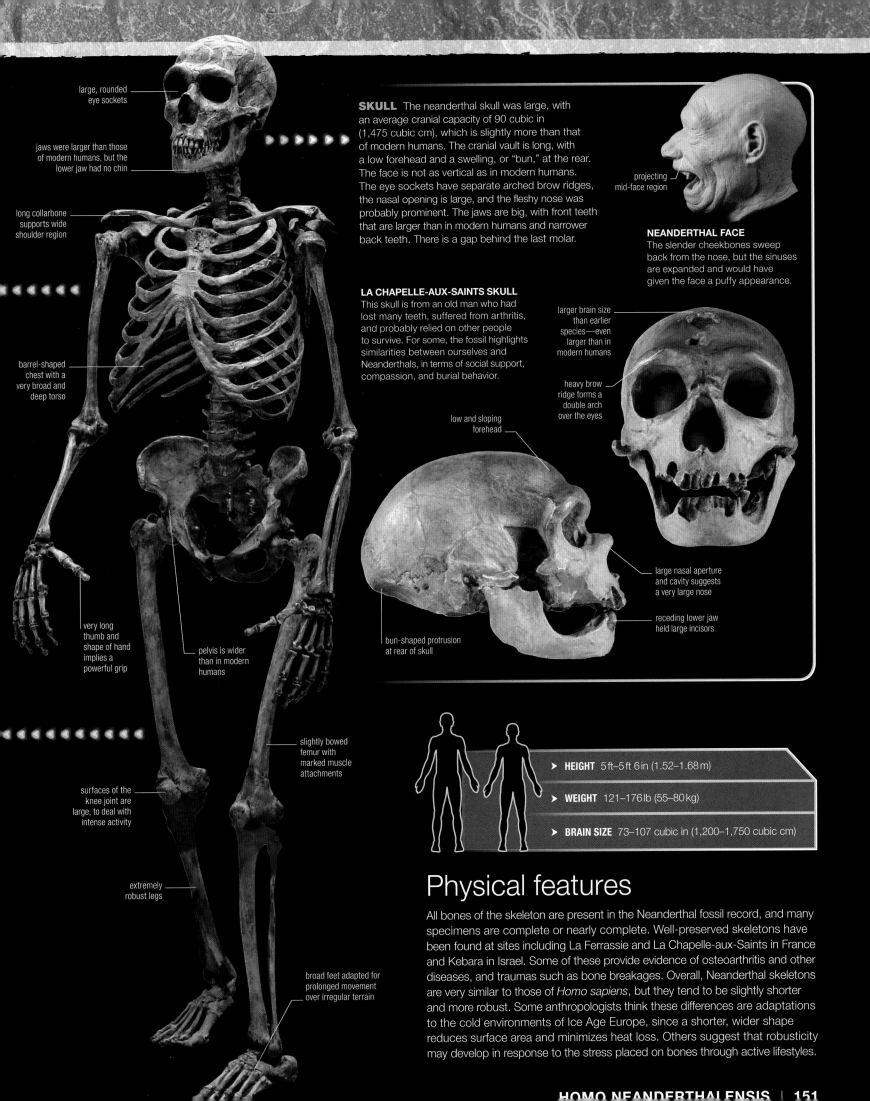

large, rounded
eye sockets

jaws were larger than those
of modern humans, but the
lower jaw had no chin

long collarbone
supports wide
shoulder region

barrel-shaped
chest with a
very broad and
deep torso

very long
thumb and
shape of hand
implies a
powerful grip

pelvis is wider
than in modern
humans

surfaces of the
knee joint are
large, to deal with
intense activity

slightly bowed
femur with
marked muscle
attachments

extremely
robust legs

broad feet adapted for
prolonged movement
over irregular terrain

SKULL The neanderthal skull was large, with an average cranial capacity of 90 cubic in (1,475 cubic cm), which is slightly more than that of modern humans. The cranial vault is long, with a low forehead and a swelling, or "bun," at the rear. The face is not as vertical as in modern humans. The eye sockets have separate arched brow ridges, the nasal opening is large, and the fleshy nose was probably prominent. The jaws are big, with front teeth that are larger than in modern humans and narrower back teeth. There is a gap behind the last molar.

projecting
mid-face region

NEANDERTHAL FACE
The slender cheekbones sweep back from the nose, but the sinuses are expanded and would have given the face a puffy appearance.

LA CHAPELLE-AUX-SAINTS SKULL
This skull is from an old man who had lost many teeth, suffered from arthritis, and probably relied on other people to survive. For some, the fossil highlights similarities between ourselves and Neanderthals, in terms of social support, compassion, and burial behavior.

larger brain size
than earlier
species—even
larger than in
modern humans

heavy brow
ridge forms a
double arch
over the eyes

low and sloping
forehead

large nasal aperture
and cavity suggests
a very large nose

receding lower jaw
held large incisors

bun-shaped protrusion
at rear of skull

▶ **HEIGHT**	5 ft–5 ft 6 in (1.52–1.68 m)
▶ **WEIGHT**	121–176 lb (55–80 kg)
▶ **BRAIN SIZE**	73–107 cubic in (1,200–1,750 cubic cm)

Physical features

All bones of the skeleton are present in the Neanderthal fossil record, and many specimens are complete or nearly complete. Well-preserved skeletons have been found at sites including La Ferrassie and La Chapelle-aux-Saints in France and Kebara in Israel. Some of these provide evidence of osteoarthritis and other diseases, and traumas such as bone breakages. Overall, Neanderthal skeletons are very similar to those of *Homo sapiens*, but they tend to be slightly shorter and more robust. Some anthropologists think these differences are adaptations to the cold environments of Ice Age Europe, since a shorter, wider shape reduces surface area and minimizes heat loss. Others suggest that robusticity may develop in response to the stress placed on bones through active lifestyles.

Archaeology

The stone-tool industry associated with *Homo neanderthalensis* in Europe is known as the Mousterian, after the site of Le Moustier in the Dordogne region of France. The Mousterian is a stone-tool industry of the Middle Paleolithic (also known as the Middle Stone Age in Africa). These Middle Paleolithic Industries, associated elsewhere with *Homo heidelbergensis* and early *Homo sapiens*, used a "prepared core" method of stone-tool knapping. The technique practiced in the Mousterian is a widespread form known as the Levallois.

Prepared cores

The Levallois technique begins with the careful selection of a raw stone nodule, and then the preparation of the nodule through knapping to produce one surface that is domed. A striking platform is created at one end which, when struck, produces a flake of a predetermined shape and size. The nodule may then be prepared again, and the process is repeated until the core is too small. This technique appears in the archaeological record between 300,000 and 200,000 years ago, and it is associated with a reduction in the number of large handaxes.

VARIETY OF TOOLS
The prepared-core technique requires skill and forward planning, is an efficient use of materials, and produces blanks that can be shaped into various forms. Traditionally, these have been described by their supposed functions, but it is just as important to consider other characteristics, such as the raw material used, the cutting-edge shape, and signs of reuse.

AWL

SCRAPER/KNIFE

HAMMER

SCRAPER/KNIFE

CORE

AXE

BURYING THE DEAD
In 1982, at Kebara Cave, Israel, the anthropologist Ofer Bar-Yosef excavated a well-preserved partial skeleton of a Neanderthal male from a depression in the ground. He was lying on his back with his arms flexed. This may be early evidence of deliberate burial.

perforated Pecten shell, painted with an orange mix of goethite and hematite

Modern behavior?

For many years, Neanderthals were portrayed as brutish, primitive, and incapable of modern behavior. However, archaeologists today are reassessing what it means to be modern and addressing the archaeological record more objectively. Some key "modern" traits identified include the ability to plan ahead, complex social networking, technological innovation, the flexibility to adapt to changing environments, symbolism, and ritual.

Consensus has been growing that Neanderthals may also have displayed many of these traits on occasion. For example, the controversial Chatelperronian industry at Arcy-sur-Cure and St. Césaire in France, thought to have been made by Neanderthals, includes innovative stone-tool types, worked bone, and items of personal adornment. Elsewhere, Neanderthals made composite tools using adhesives (Harz Mountains, Germany), hunted birds for feathers and exploited aquatic resources (Vanguard and Gorham's Caves, Gibraltar), hunted selectively (Ortvale Kida Rockshelter, Georgia), and buried their dead (La Ferrassie, France). Neanderthals were clearly capable of very complex behaviors, suggesting that the cognitive abilities displayed by *Homo sapiens* did not suddenly appear in a revolutionary biological change, but have a long evolutionary history.

NEANDERTHAL "MAKE-UP"
Perforated shells partly covered with the residue of pigments have been found in Spanish caves inhabited by Neanderthals. The archaeologist João Zilhao suggests that they may have been dishes to hold dyes for body decoration or coloring hides.

Lifestyle and hunting

Neanderthals occupied Europe for at least 100,000 years during a period when the climate was dominated by cold glacial cycles. For many Neanderthals, life in the Ice Age would have been harsh, but they were clearly very resourceful and successful.

The wide range of habitats in which Mousterian sites are found, from coasts to uplands, and the targetnig of different animals at different times of year in highly seasonal environments (often reindeer in winter and red deer in summer) suggest Neanderthals were highly adaptable hunters. Chemical analysis of fossil bone suggests meat was an important part of Neanderthals' diets, but they also exploited freshwater and marine fish and shellfish and marine mammals, and recent study of dental plaque has revealed that they ate a range of plant foods, including pine nuts and other seeds, moss, mushrooms, and tubers. At some sites (including Moula-Guercy, France), it is thought that Neanderthals may have practiced cannibalism: the hominin bones here show the same patterns of breaks, stone-tool cut-marks, and distribution as butchered deer. However, such cannibalism may have been practiced for ritual rather than primarily nutritional reasons.

IBEX PREY
Neanderthals were clearly skilled hunters who were well able to target a range of species, including ibex. These mountain goats live in inaccessible areas and require considerable planning and skill to hunt successfully.

Skull was fractured diagonally, probably by a sharp weapon

NEANDERTHAL INJURIES
Many Neanderthal fossils, including this skull from Saint-Césaire, France, bear evidence of violent injuries. A reliance on handheld spears obliged them to get very close to large wild animals to make a kill. Later, *Homo sapiens* invented thrown spears and eventually bows and arrows, which may have made hunting a slightly safer occupation.

NEANDERTHAL MEDICATION
Some plants consumed by Neanderthals had medical properties, including penicillin fungus, poplar (which contains the active ingredient in aspirin), chamomile (shown here), and yarrow. Did they deliberately use these plants to treat illness?

PIT OF BONES

Since 1997, excavations at La Sima de los Huesos ("Pit of Bones") in Spain have revealed more than 5,500 fossil hominin remains. These were originally assigned to *Homo heidelbergensis* but also show some unique Neanderthal traits, suggesting these individuals represented a late *H. heidelbergensis* group in the process of evolving into their descendants, the Neanderthals. However, DNA recovered from the bones recently demonstrated that they are more closely related to Neanderthals. The site is a 43ft- (13m-) deep vertical shaft accessible only by 1,968ft (600m) of difficult passages. The bones were found jumbled at the base, leading some anthropologists to believe that bodies were deliberately thrown down, perhaps as a symbolic "burial."

RITUAL DEPOSIT
Made of pink quartzite, this unused handaxe is the only stone tool found in association with the Sima de los Huesos hominin fossils. Some archaeologists believe it is evidence of ritual burial.

Homo *neanderthalensis:*
Some of the last surviving Neanderthals collect shellfish from a lagoon located near the Gibraltar headland.

FAMILY UNITS
While most experts believe that Neanderthals lived in strong family groups, new genetic evidence suggests that these groups had what is known as a "patrilocal" base. This means that while males born into a family would have stayed with the group into adulthood, females would move away into other families.

BODY ART
There is growing evidence to suggest that Neanderthals may have worn ornaments, such as pierced shells and feathers, and perhaps even used body art. Since some scientists view body art and adornment as a form of communication, this hints at the possibility that Neanderthals used language of some kind.

POSSIBLE CLOTHING
Although no traces of Neanderthal clothing have survived, it is inconceivable to think that they would not have attempted to protect their bodies from the European elements. There is no evidence of Neanderthals using needles—unlike early modern humans—but they probably fashioned animal hides into wrap-around garments.

GIBRALTAR CAVES
Hearths and stone tools revealed by excavations at Gorham's Cave in Gibraltar have revealed evidence of what may have been some of the last Neanderthals in Europe, dating to perhaps as late as 28,000 years ago. There is also evidence of what could be abstract "art" —a geometric grid inscribed on the cave wall.

RED HAIR, PALE SKIN
Scientists studying Neanderthal DNA have discovered that some Neanderthals would have had red hair and pale skin, and that hair and skin pigments might have been as varied as those of modern Europeans. Their bodies may have evolved pale skin to help with the synthesis of vitamin D in areas of lower sun exposure.

LOW SEA LEVEL
At the time that Neanderthals inhabited this area, the sea level was much lower, due to glacial conditions in Northern Europe. What is now Gibraltar would have been surrounded by a coastal plain of sand dunes, marshes, and lagoons, teeming with animal life that presented an abundance of food options.

BROAD DIET
Evidence from Vanguard and Gorham's Caves has shown that Neanderthals enjoyed a much broader diet than was previously supposed. Easy access to the water enabled them to capture monk seals, while they ate dolphins, which probably washed up on the shore. There is also evidence of them cooking mussels to open the shells.

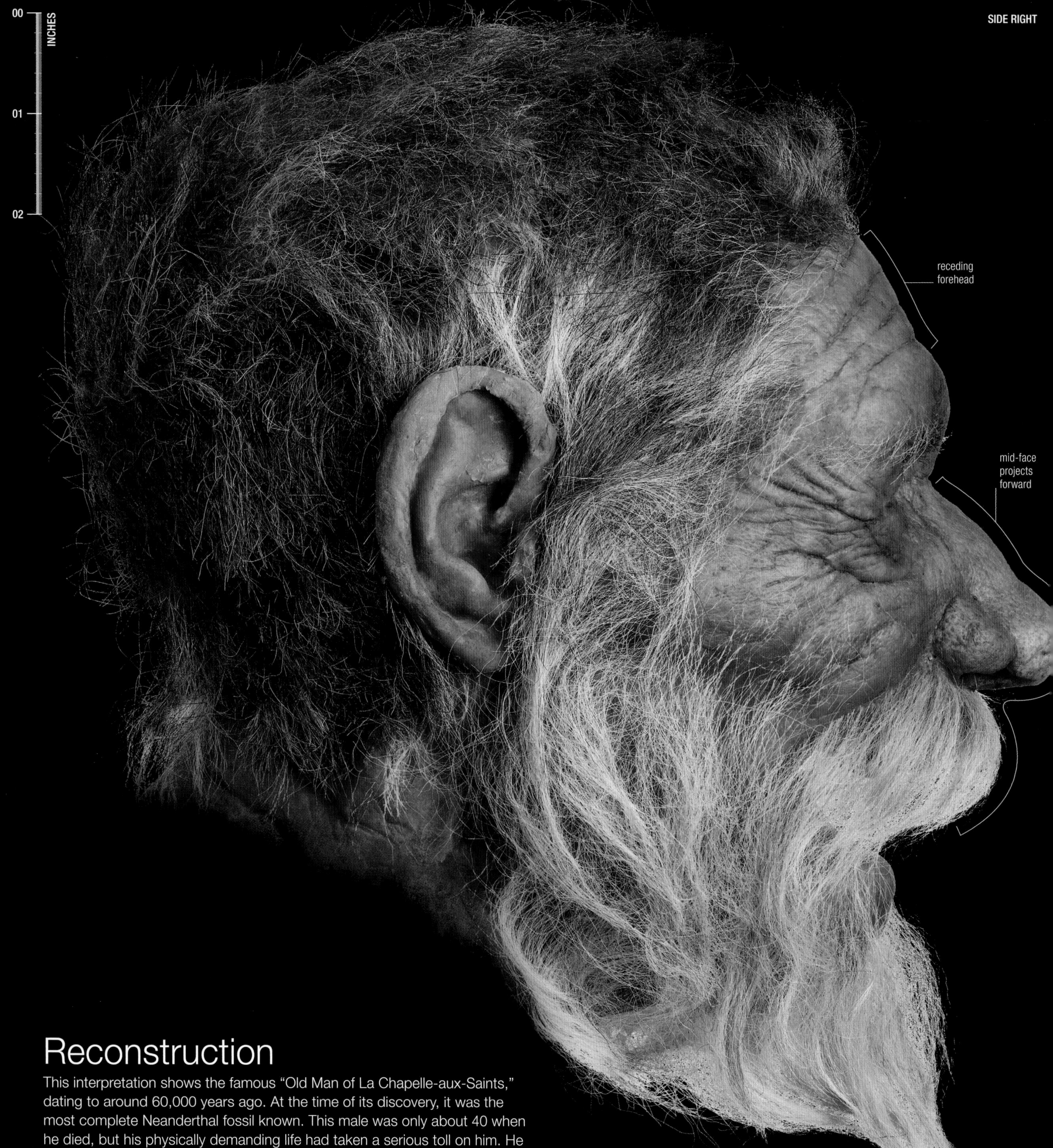

INCHES

00

01

02

receding forehead

mid-face projects forward

Reconstruction

This interpretation shows the famous "Old Man of La Chapelle-aux-Saints," dating to around 60,000 years ago. At the time of its discovery, it was the most complete Neanderthal fossil known. This male was only about 40 when he died, but his physically demanding life had taken a serious toll on him. He suffered from widespread degeneration and arthritis in his joints, especially his left hip, and a number of healed wounds, including a fractured rib. Such features resulted in early reconstructions of this man as a shuffling, slouched figure, contributing to a false idea of Neanderthals as brutish cavemen.

FRONT RIGHT

Neanderthals probably had paler skin than their African ancestors, perhaps in response to the relatively lower levels of light in the northerly latitudes of their Eurasian range

forward-placed jaws and large front teeth, may have been used to grip meat, hide, or sinew while they were being cut

FRONT

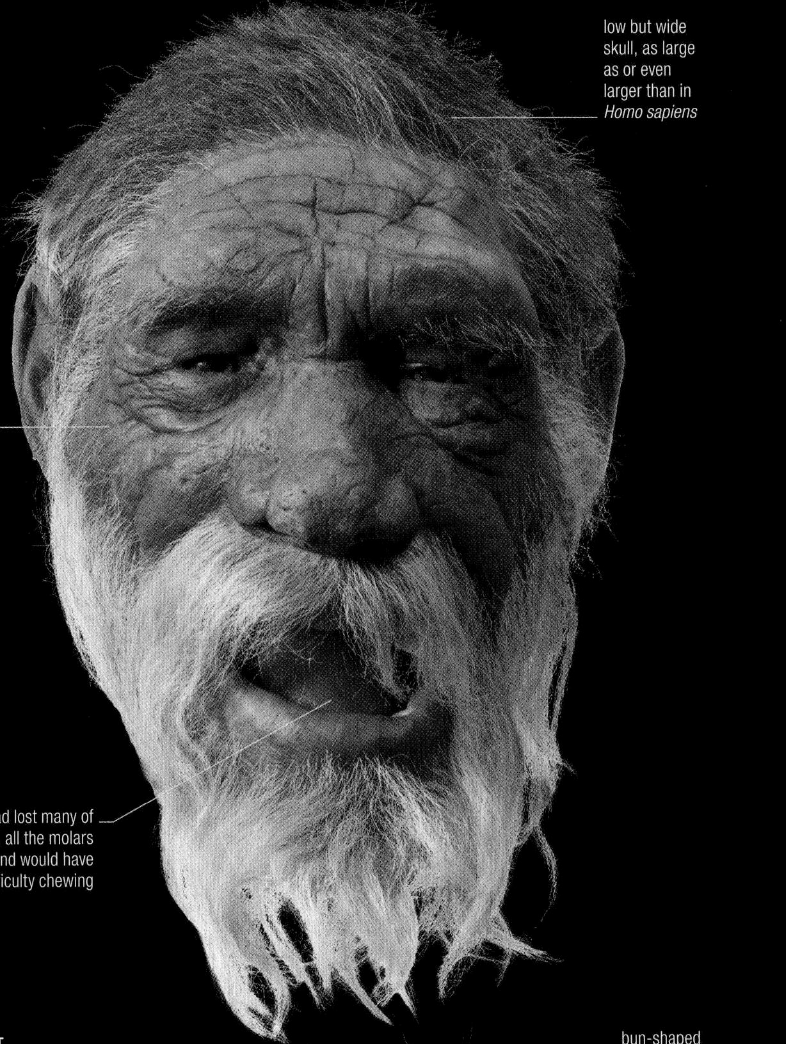

low but wide skull, as large as or even larger than in *Homo sapiens*

broad and backward-sloping cheekbones

this individual had lost many of his teeth, including all the molars in his lower jaw, and would have had difficulty chewing

FRONT LEFT

heavy brow ridge over each eye forming a double arch

SIDE LEFT

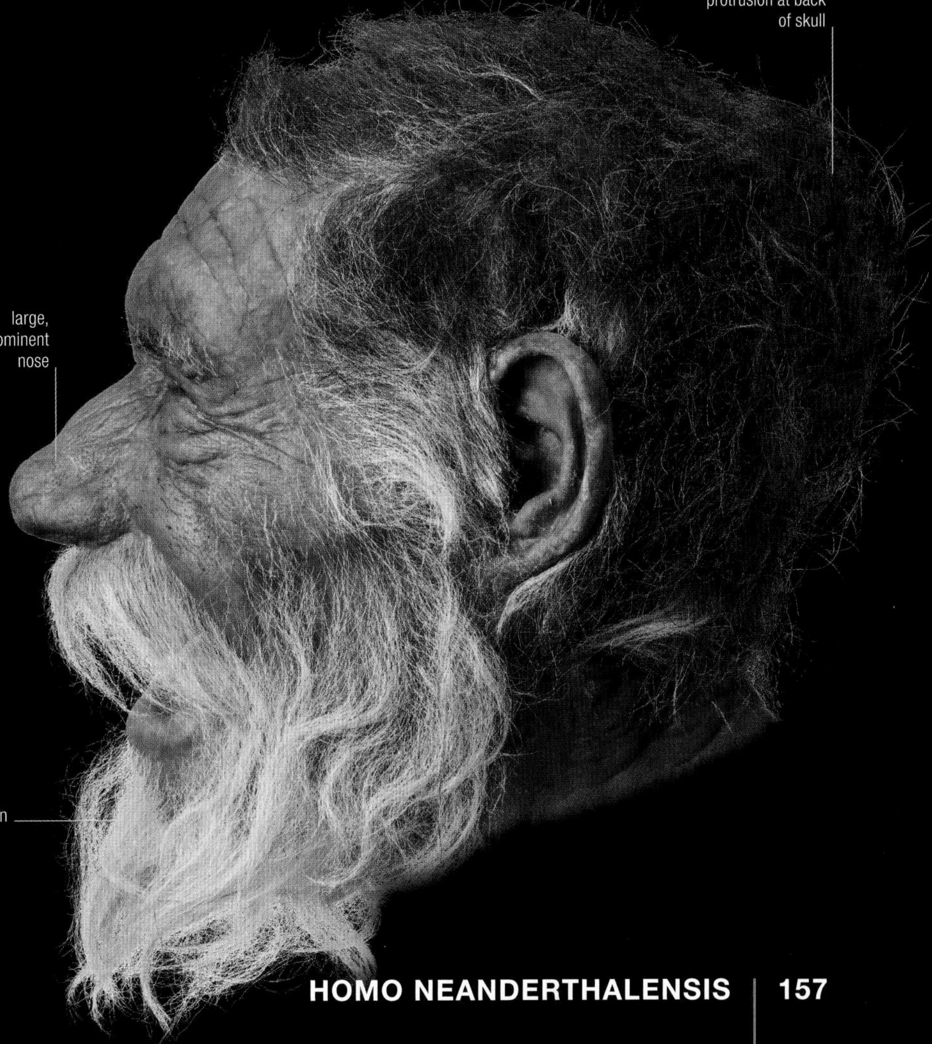

bun-shaped protrusion at back of skull

large, prominent nose

little or no chin

HOMO NEANDERTHALENSIS | 157

Homo naledi

***Homo naledi* is a recent addition to the human family tree. Its unusual combination of old and new traits is all the more remarkable since it has been dated to very recently in the timeline of human evolution.**

Discovery

The Rising Star cave system in Gauteng, South Africa, was first investigated for anthropological remains in 2013, after cavers discovered the Dinaledi chamber at the bottom of a 39ft (12m) shaft and spotted fossil bones covering its floor. Photos taken by the cavers were given to anthropologist Lee Berger, who led the Rising Star Expedition to excavate the Dinaledi and

nearby Lesedi chambers later that year. In 2015, the remains were formally assigned to a new species of the genus *Homo*. The process of excavation and publication of results attracted controversy, as Berger and his team of relatively young, early career researchers took only two years to excavate and publish their findings. Furthermore, rather than publishing them in a traditional

science journal, they released them online, in an open-access journal, and made 3-D scans of the fossils available for anyone to download and print. Critics believe that the publicity surrounding the excavation drove a rush to publish in a much shorter time frame than is usual, which may have compromised the analysis of the bones and prevented sufficiently in-depth study and review by other scientists. However, others believe that this innovative approach makes science much more open and democratic.

DID HOMO NALEDI BURY ITS DEAD?

How did the remains reach these inaccessible caves? There is no evidence of predators or scavengers, or of flowing water that might have carried the bones in, and the fossils represent several individuals of all ages that appear to have accumulated over some time. Some anthropologists believe that the bodies may have been intentionally disposed of here in some kind of burial ritual—implying an "advanced" behavior. However, others argue that the shaft simply provided a way of disposing of decaying corpses.

ABOVE-GROUND CONTROL
With only limited numbers of people able to reach and work in the cramped conditions of the Dinaledi chamber, other experts could only watch the excavations via camera.

HOMO NALEDI

➤ **NAME MEANING** "Star man"

➤ **LOCALITY** The Dinaledi and Lesedi chambers in the Rising Star cave system, Gauteng, South Africa

8 MYA	PRESENT	**AGE** *335,000–236,000 years ago*
7 MYA	1 MYA	Dated using Carbon-14, optically stimulated
6 MYA	2 MYA	luminescence (OSL), paleomagnetism,
5 MYA	3 MYA	uranium-series, and electron spin resonance
4 MYA		

➤ **FOSSIL RECORD** Fragmentary, disarticulated and articulated remains of almost the full skeleton from at least 18 individuals of a range of ages, including 5 skulls

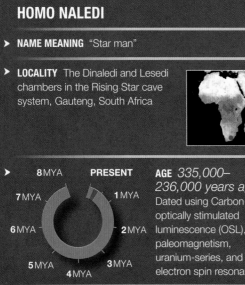

UNDERGROUND ASTRONAUTS
To reach the fossils, excavators had to be small enough to squeeze through passages as narrow as 7in (18cm) wide. The six women—who were both scientists and cavers—that were recruited were nicknamed "underground astronauts."

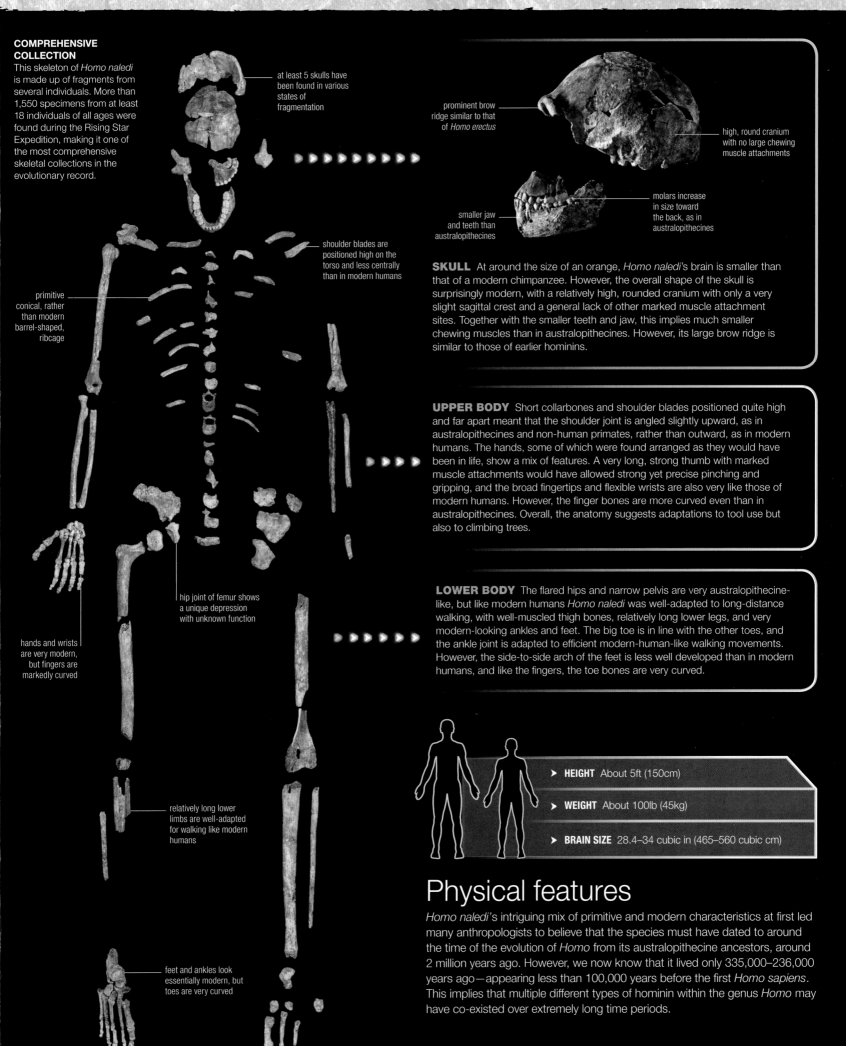

COMPREHENSIVE COLLECTION

This skeleton of *Homo naledi* is made up of fragments from several individuals. More than 1,550 specimens from at least 18 individuals of all ages were found during the Rising Star Expedition, making it one of the most comprehensive skeletal collections in the evolutionary record.

at least 5 skulls have been found in various states of fragmentation

shoulder blades are positioned high on the torso and less centrally than in modern humans

primitive conical, rather than modern barrel-shaped, ribcage

hip joint of femur shows a unique depression with unknown function

hands and wrists are very modern, but fingers are markedly curved

relatively long lower limbs are well-adapted for walking like modern humans

feet and ankles look essentially modern, but toes are very curved

prominent brow ridge similar to that of *Homo erectus*

high, round cranium with no large chewing muscle attachments

smaller jaw and teeth than australopithecines

molars increase in size toward the back, as in australopithecines

SKULL At around the size of an orange, *Homo naledi*'s brain is smaller than that of a modern chimpanzee. However, the overall shape of the skull is surprisingly modern, with a relatively high, rounded cranium with only a very slight sagittal crest and a general lack of other marked muscle attachment sites. Together with the smaller teeth and jaw, this implies much smaller chewing muscles than in australopithecines. However, its large brow ridge is similar to those of earlier hominins.

UPPER BODY Short collarbones and shoulder blades positioned quite high and far apart meant that the shoulder joint is angled slightly upward, as in australopithecines and non-human primates, rather than outward, as in modern humans. The hands, some of which were found arranged as they would have been in life, show a mix of features. A very long, strong thumb with marked muscle attachments would have allowed strong yet precise pinching and gripping, and the broad fingertips and flexible wrists are also very like those of modern humans. However, the finger bones are more curved even than in australopithecines. Overall, the anatomy suggests adaptations to tool use but also to climbing trees.

LOWER BODY The flared hips and narrow pelvis are very australopithecine-like, but like modern humans *Homo naledi* was well-adapted to long-distance walking, with well-muscled thigh bones, relatively long lower legs, and very modern-looking ankles and feet. The big toe is in line with the other toes, and the ankle joint is adapted to efficient modern-human-like walking movements. However, the side-to-side arch of the feet is less well developed than in modern humans, and like the fingers, the toe bones are very curved.

> **HEIGHT** About 5ft (150cm)

> **WEIGHT** About 100lb (45kg)

> **BRAIN SIZE** 28.4–34 cubic in (465–560 cubic cm)

Physical features

Homo naledi's intriguing mix of primitive and modern characteristics at first led many anthropologists to believe that the species must have dated to around the time of the evolution of *Homo* from its australopithecine ancestors, around 2 million years ago. However, we now know that it lived only 335,000–236,000 years ago—appearing less than 100,000 years before the first *Homo sapiens*. This implies that multiple different types of hominin within the genus *Homo* may have co-existed over extremely long time periods.

Homo *sapiens*

The earliest members of the species *Homo sapiens*
represented the evolutionary transition from African
Homo heidelbergensis to the first modern humans.

After more than 7 million years of hominin evolution, *Homo sapiens* is the only surviving branch of the hominin family tree.

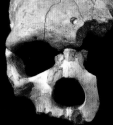
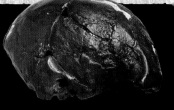

Homo sapiens appeared in Africa around 300,000 years ago. Today *Homo sapiens* is the last living hominin species, and populations have expanded into every corner of the world. Debate rages, however, as to why we survived while all other hominin species became extinct. Which species did *Homo sapiens* encounter as they spread? Did they displace them or interbreed with them?

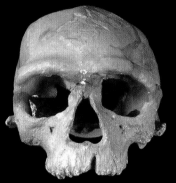

FLORISBAD
This partial skull from South Africa has been controversial because the low, broad skull and large face do not fit easily within the range of modern human morphology.

OMO II
Until recently, the two fossil skulls excavated from Omo Kibish, Ethiopia, dating to 195,000 years ago, represented the oldest known *Homo sapiens* remains.

The first modern humans

The origins of modern humans were long thought to lie in the Rift Valley of East Africa around 200,000 years ago. However, recent redating of Jebel Irhoud in Morocco has established that the fossils here are 300,000 years old, making them the earliest to show clearly modern characteristics such as short, flat faces. All of the early African fossils are quite variable and show a range of modern characteristics alongside ancestral traits, such as robust brow ridges and long, low braincases rather than the high, rounded ones of modern humans. For example, the two individuals from the Omo Basin in Ethiopia both date to 195,000, but one (Omo I) looks much more modern than the other (Omo II). Other early modern fossils include those from Herto in Ethiopia (155,000 years old), Singa in Sudan (133,000 years old), Laetoli in Tanzania (120,000 years old), and Border Cave and Klasies River Mouth in South Africa (120,000–90,000 years old). By 100,000 years ago, modern humans had expanded into western Asia: more than 20 fully modern individuals are buried at Skhul and Qafzeh in Israel (120,000–80,000 years old).

JEBEL IRHOUD
This cave site in Morocco has recently been redated to 300,000 years old, making the fossil found here the oldest known *Homo sapiens* specimen.

HOMO SAPIENS IDALTU
Some anthropologists suggest that the intermediate morphology of the 155,000-year-old crania from Herto, Ethiopia, indicates a new subspecies.

OMO RIVER VALLEY
In the late 1960s to early 1970s, fieldwork at the fossil-bearing deposit known as the Kibish Formation, in the Omo River Valley of Ethiopia, yielded the fragmented skull and skeleton remains of two early modern humans.

HOMO SAPIENS

> **NAME MEANING** "Wise man"

> **LOCALITY** Various sites across Africa, then spreading worldwide

> 8 MYA   PRESENT   **AGE** *300,000 years ago to present day*
> 7 MYA   1 MYA   Dated through various techniques, including
> 6 MYA   2 MYA   radiocarbon dating
> 5 MYA   3 MYA
>   4 MYA

> **FOSSIL RECORD** Complete skulls; also various partial or fragmented skulls and skeletons

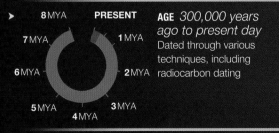

QAFZEH 6 SKULL

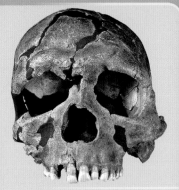

The cave site of Jebel Qafzeh in Israel contained more than a dozen human burials dating to 90,000 years ago. Two are particularly well preserved. Qafzeh 9 is an adult buried in a flexed position along with the body of an infant. Qafzeh 6 has a well-preserved skull, possibly of a young male. The morphology is varied, with one group (including Qafzeh 6) showing a strong brow ridge running continuously across the face.

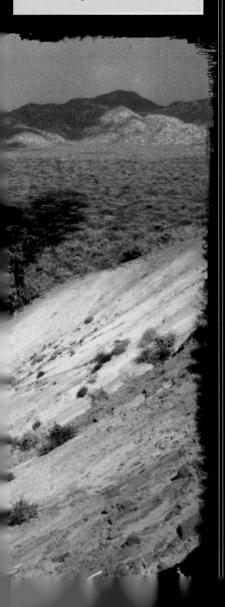

SKULL The *Homo sapiens* skull has many functions, including accommodating a large brain, housing the ears and eyes, and anchoring the facial muscles and chewing apparatus—all while balancing on a vertical spine. Many features typically described as modern relate to these functions and how they work in combination.

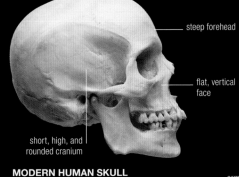

steep forehead

flat, vertical face

short, high, and rounded cranium

MODERN HUMAN SKULL
The cranium is high, has thin walls, and a vertical frontal bone. It is also rounded, with weak muscle markings and a small, flat face tucked underneath.

Physical features

The skeleton of *Homo sapiens*, in common with other more recent hominins, shows adaptations for efficient bipedalism, but is distinctive in being lightly built. The lower limbs are long and gracile (slender), and the thigh bones are angled to keep the center of gravity tight under the body (see p.69). The pelvis is quite narrow and short, particularly in males, and the weight-bearing hip-joint surface is large. In the foot, the toes are stocky, the arches are well developed, and the ankle bones are big and strong. The spine is curved to maintain balance and distribute body weight effectively during walking, with large vertebrae in the lower (lumbar) region. The rib cage is barrel shaped, the arms are relatively short, and the hands are dextrous, with opposable thumbs and excellent precision grip. Modern humans have very large brains for their body size, averaging 79 cubic in (1,300 cubic cm).

MODERN HUMAN VARIATION
Living populations of *Homo sapiens* show moderate skeletal variation. Body size and robusticity is slightly sexually dimorphic and may have deep evolutionary roots, but it is likely that ethnic or geographic variants are relatively recent.

> **HEIGHT** 5–6 ft (1.5–1.8 m)

> **WEIGHT** 120–183 lb (54–83 kg)

> **BRAIN SIZE** 61–122 cubic in (1,000–2,000 cubic cm)

small teeth, e
front teeth

protruding chin

shoulder bla
and position
on back of ri

narrow shoulders

arms short relative to leg length, compared to early hominins

short and thick pubic bone

small, long fingertips relative to those of Neanderthals

long
with
attac

long, slender shin bones with relatively thin walls

huma
slightl
than t
Neanc
sugge
that h
have k
to run
and fa

big toe aligned with rest of toes, rather than divergent

Archaeology

Distinctive modern skeletal traits first appear in the African fossil record around 300,000 years ago, but it is less clear when and where modern behavior arose. Much debate surrounds the definition of behavioral modernity. Key characteristics might include the ability to plan ahead, complex social networking, technological innovation, the flexibility to adapt to changing environments, and the use of symbolism and ritual. It is still unclear if these traits appear before or after the evolution of physically modern people.

Refined tools

By 200,000 years ago, many innovations had been made in stone tool technology. Large handaxes became less common, and were replaced with a range of smaller tools in more diverse toolkits. Tools made of flakes were preferred over large cores, and production was more efficient, with improvements in the amount of cutting edge produced on a raw nodule. Great skill and craftsmanship was required to create these standardized tools. One of the key innovations was the introduction of core preparation, which allowed flakes of a predetermined size and shape to be struck off with a single blow. Some flakes were small, and retouched into points for attaching to spears or setting into shafts with glue. Others were long and thin, perhaps used for piercing skins or woodworking. This period of stone tool technology persisted until relatively recently, at least 50,000 years ago, and is known in Africa as the Middle Stone Age.

STANDARDIZED TOOLS
In the African Middle Stone Age, toolmakers were skilled at producing standardized and symmetrical artifacts with straight, sharp cutting edges, fine points, and blunt ends for holding or hafting.

Modern behavior

One of the reasons why the first appearance of modern behavior is difficult to identify is that some key characteristics, such as technological innovation, are more visible in the archaeological record than others, such as ritual. However, in Africa there is growing archaeological evidence of early modernity. By at least 120,000 years ago, people were transporting stone tools long distances across the landscape, indicating trade or seasonal migrations. Toolkits also begin to include small, lighter forms, bone points for fishing, and hafting technologies. Occupation sites become more organized, with clear separation of activities such as cooking and burying the dead. The most overwhelming evidence for modern human behavior, however, comes much later from Europe, at about 40,000 years ago, when cave art and sculpture provide unambiguous evidence for use of symbolism.

BLOMBOS CAVE
A collaborative excavation at Blombos Cave, South Africa, between South African anthropologist Christopher Henshilwood and the Iziko South African Museum has revealed a wealth of evidence for early modern human activity including hearths, bone tools, fishing, and ocher.

RED OCHER
Ocher is an iron oxide that may have been used as body paint or for decorating artifacts, and is sometimes found in burials. Two pieces of engraved ocher, at least 70,000 years old, were found near the hearth at Blombos Cave.

arm bent at elbow

skull is lying on
its side

deer antlers placed
across upper body

RITUAL BURIAL
This skeleton of an adolescent early *Homo sapiens* was found at the Qafzeh Cave site near Nazareth, Israel, buried in a pit dug into bedrock. The skeleton was lying on its back with both arms flexed upward and a set of deer antlers laid across the chest. The burial dates to around 100,000–90,000 years ago and may represent a form of ritual, possibly indicating a belief in an afterlife.

left arm
just visible

INCHES

00

01

02

skull still quite long and
wide, and the vault relatively
low, as in older hominins

relatively
strong
forehead

profile of face
very modern,
being flat and
with little
projection of
the jaw

Reconstruction

This reconstruction is based on one of the earliest known fossils of
Homo sapiens—a nearly complete skull discovered along with other
fossils in 1961 at the site of Jebel Irhoud, near Marrakech in Morocco.
The fossils, dated to around 300,000 years ago, were initially thought
to be African relatives of the Neanderthals. In fact, they are older than
many of the classic European Neanderthals. It is now known that the
"archaic" features they display are also shared by other early *Homo
sapiens* from North and East Africa and the Near East.

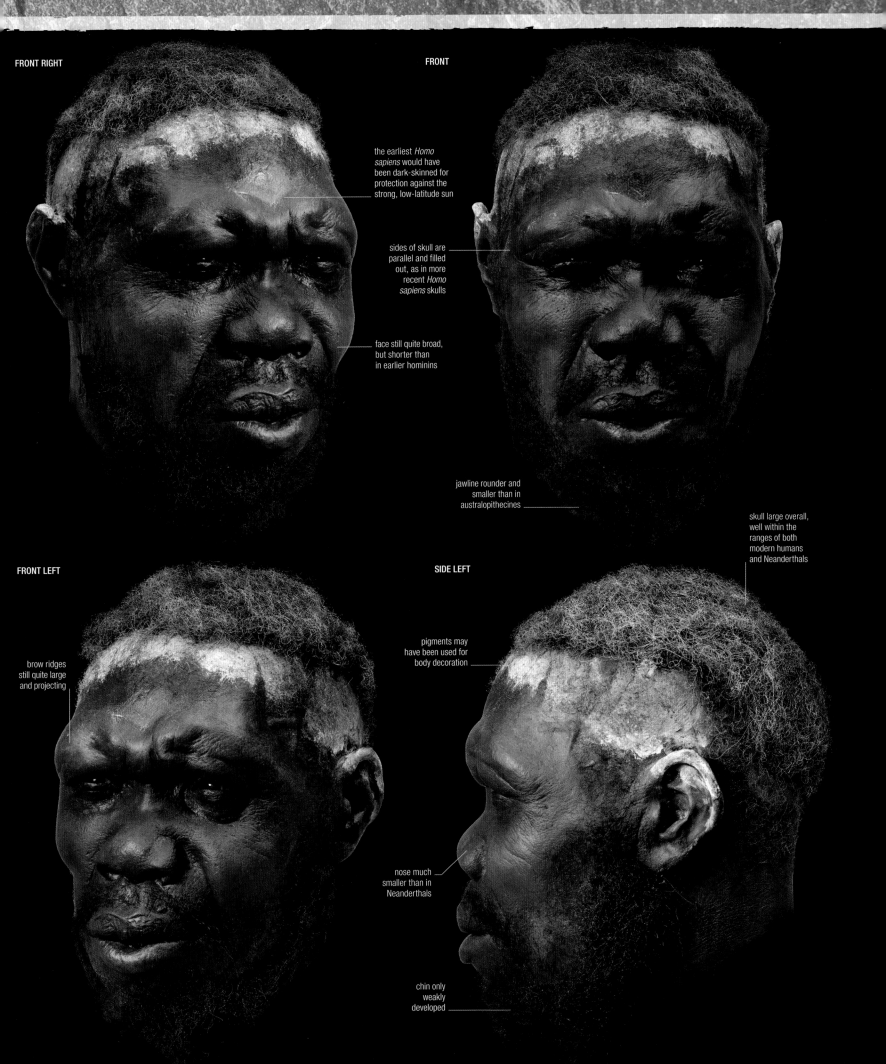

FRONT RIGHT

FRONT

the earliest *Homo sapiens* would have been dark-skinned for protection against the strong, low-latitude sun

sides of skull are parallel and filled out, as in more recent *Homo sapiens* skulls

face still quite broad, but shorter than in earlier hominins

jawline rounder and smaller than in australopithecines

skull large overall, well within the ranges of both modern humans and Neanderthals

FRONT LEFT

SIDE LEFT

brow ridges still quite large and projecting

pigments may have been used for body decoration

nose much smaller than in Neanderthals

chin only weakly developed

high, bulging
forehead may
indicate some
reorganization of
the brain within

ve
tu
u
b

distinct
chin

Reconstruction

This young woman, who died at about the age of 20, was buried at the
cave site of Qafzeh, Israel. In all, 21 skeletons representing a variety of ages
were found here, including that of a very young child found buried at the
woman's feet. The woman's skeleton—one of the most complete—was
recognized as that of a modern human, albeit one with some fairly "archaic"
features. In the 1980s, the Qafzeh site was reliably dated to around 100,000
years ago—almost twice as old as previous estimates—showing that her
skeleton was one of the earliest examples of modern humans in the region

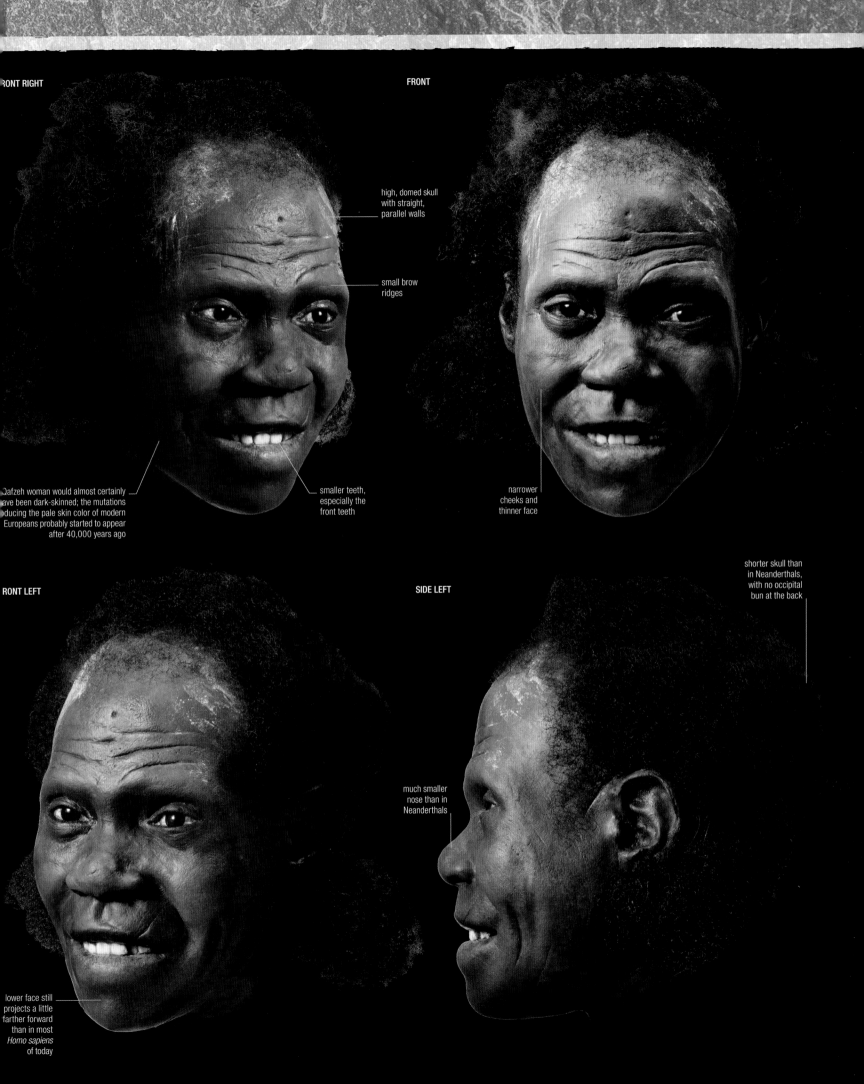

FRONT

high, domed skull
with straight,
parallel walls

small brow
ridges

Qafzeh woman would almost certainly
have been dark-skinned; the mutations
producing the pale skin color of modern
Europeans probably started to appear
after 40,000 years ago

smaller teeth,
especially the
front teeth

narrower
cheeks and
thinner face

shorter skull than
in Neanderthals,
with no occipital
bun at the back

FRONT LEFT

SIDE LEFT

much smaller
nose than in
Neanderthals

lower face still
projects a little
farther forward
than in most
Homo sapiens
of today

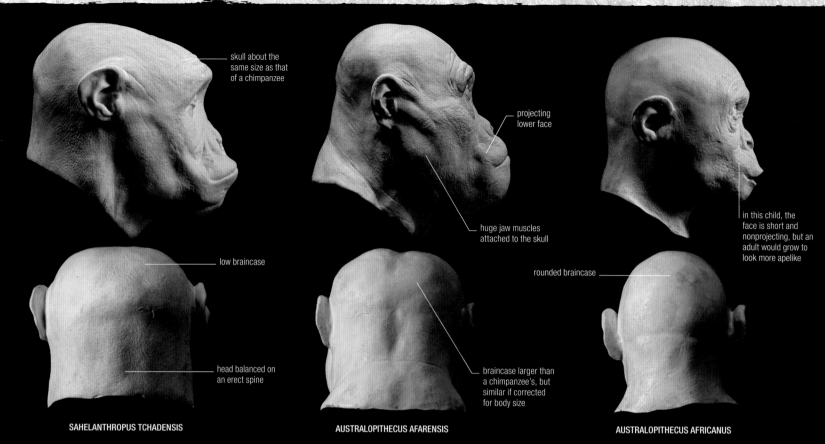

skull about the same size as that of a chimpanzee

projecting lower face

huge jaw muscles attached to the skull

in this child, the face is short and nonprojecting, but an adult would grow to look more apelike

low braincase

rounded braincase

head balanced on an erect spine

braincase larger than a chimpanzee's, but similar if corrected for body size

SAHELANTHROPUS TCHADENSIS

AUSTRALOPITHECUS AFARENSIS

AUSTRALOPITHECUS AFRICANUS

Comparing heads

The physical differences between the earliest and most recent hominins are characterized by changes in the size, shape, and proportions of the different regions of the body. This is particularly clear in the head.

The very first hominins displayed many features that reflected their Miocene ape ancestry. *Sahelanthropus* had a brain the size of a chimpanzee's—one third the size of modern human's. This was contained within a low, narrow braincase that had a heavy brow ridge and sloping forehead. Its long jaws (containing large back teeth) and flat nose gave its face a prognatic, or projecting, profile.

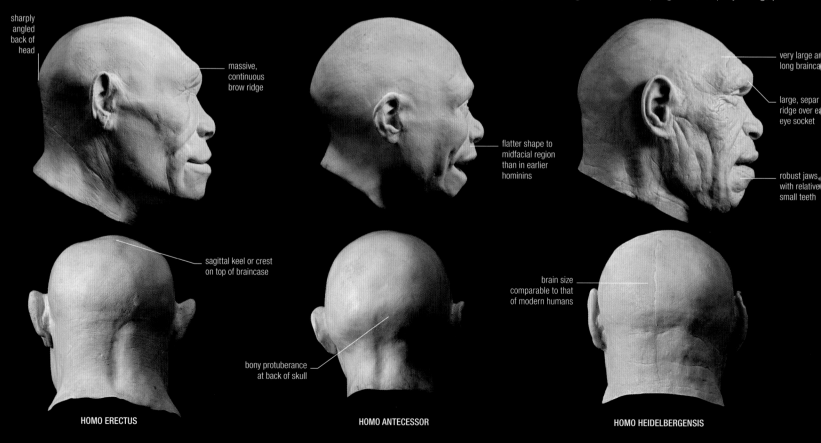

sharply angled back of head

massive, continuous brow ridge

very large ar long brainca

large, separ ridge over e eye socket

flatter shape to midfacial region than in earlier hominins

robust jaws with relative small teeth

sagittal keel or crest on top of braincase

brain size comparable to that of modern humans

bony protuberance at back of skull

HOMO ERECTUS

HOMO ANTECESSOR

HOMO HEIDELBERGENSIS

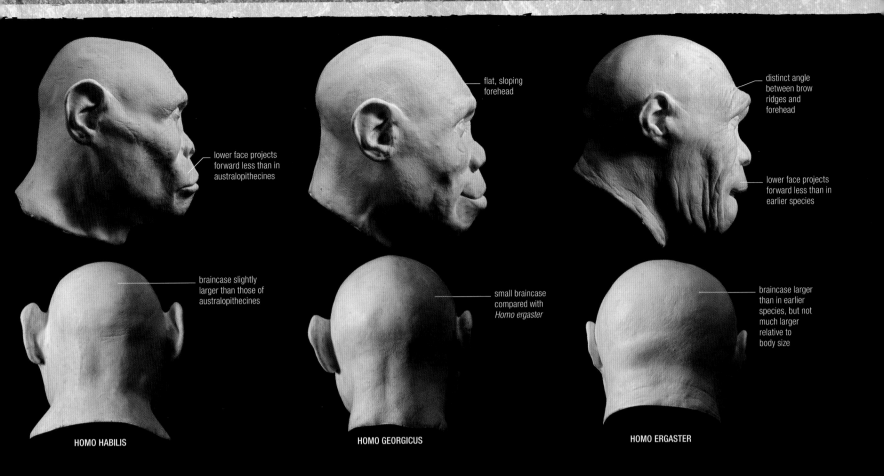

lower face projects forward less than in australopithecines

flat, sloping forehead

distinct angle between brow ridges and forehead

lower face projects forward less than in earlier species

braincase slightly larger than those of australopithecines

small braincase compared with *Homo ergaster*

braincase larger than in earlier species, but not much larger relative to body size

HOMO HABILIS

HOMO GEORGICUS

HOMO ERGASTER

While the australopithecines also had a projecting face—with particularly robust jaws and teeth—some species had a rounded braincase and a sloping forehead. However, their brains were also very small—around 24–31 cu in (400–500 cc)—compared to later species. One of the key features that defined *Homo habilis* as the first member of a new genus was the size of its brain—at around 36–43 cu in (600–700 cc) it marked a significant increase. With subsequent *Homo* species, the jaws became shorter, making the face more vertical and less projecting, while the braincase generally grew in size and became high and domelike. The familiar profile of *Homo sapiens* includes one feature present in no other hominin—a chin.

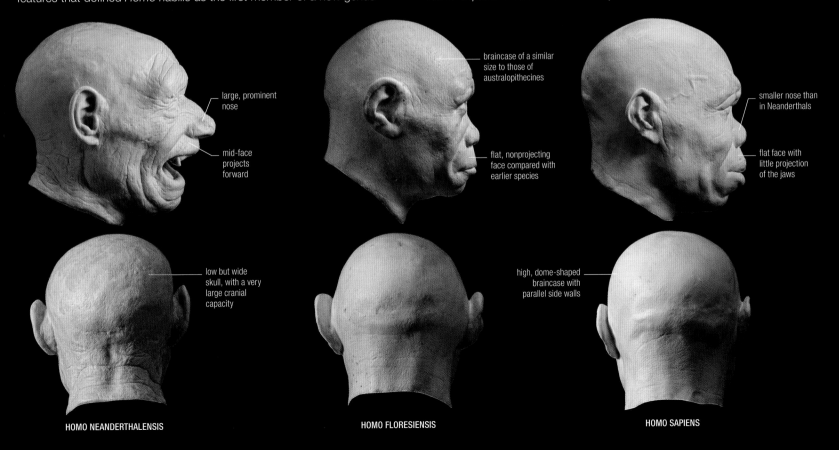

large, prominent nose

mid-face projects forward

braincase of a similar size to those of australopithecines

flat, nonprojecting face compared with earlier species

smaller nose than in Neanderthals

flat face with little projection of the jaws

low but wide skull, with a very large cranial capacity

high, dome-shaped braincase with parallel side walls

HOMO NEANDERTHALENSIS

HOMO FLORESIENSIS

HOMO SAPIENS

OUT OF AFRICA

Homo sapiens has proven to be a very successful
species of primate. We have not adapted to one
particular ecological niche, but rather, our ancestors
evolved to be increasingly adaptable. We can survive
by eating a huge variety of different foods, and our
flexible behavior—which includes the use of culture
and technology—has enabled us to survive in a wide
range of different environments, from the Arctic to the
tropics. However, we may now be reaching a critical

Human migrations

We are a truly global species. There are humans just about everywhere—on every continent, on almost every scrap of land. But our ubiquity is no recent phenomenon. The colonization of the entire planet by *Homo sapiens* started more than 50,000 years ago.

Episodes of expansion

Ours was not the first human species to emerge from its African homeland. While early hominins appear to have been exclusively African species, at least one other human species emerged from Africa into Asia before our own ancestors made that journey. A story of the long-legged *Homo ergaster* evolving in Africa then emerging nearly 2 million years ago to spread across Asia has become widely accepted. This story fits with the appearance of *Homo erectus* fossils in East Asia, dating to around 1 million years ago. But surprising fossil discoveries, including those at Dmanisi in Georgia and on the Indonesian island of Flores, have recently provoked researchers to suggest that other hominin species may also have made that journey out of Africa, much earlier than previously thought. The picture will become clearer as more fossil and archaeological evidence comes to light.

MIGRATION ROUTES
Some time around 100,000 years ago, modern humans started to leave Africa, and, generation by generation, they gradually spread across the globe. The arrows on this map represent hypothetical migration and colonization routes based on archaeological and genetic evidence (arrows also in part based on a map by Stephen Oppenheimer, 2008).

BARRIER TO EXPANSION

For long periods during the Pleistocene, the way out of Africa would have been blocked. During glacial periods, the climate in Africa was cold and dry, with the Sahara and Sinai deserts spreading to form an impassable barrier. However, at intervals of about 100,000 years, the climate became warmer and wetter, turning parts of the desert green, and allowing migrations out of the African continent.

KEY

→ POSSIBLE MIGRATION ROUTES OF HOMO SAPIENS

● SITE OF EARLY HOMO SAPIENS

Beringia L.. Bridge

Swan Point
Bluefish Caves
Tuluaq Hill (Sluiceway-Tuluaq comple...

BERINGIA
The Bering Straits from northeastern Siberia to Alaska are an obvious colonization route. In the last ice age this area would have been dry land, a vast plain known as Beringia.

N O R T H   A M E R I C A

Clovis

15,000 YEARS AGO

Meadowcroft

Cactus Hill

E U R O P E

45,00.. YEARS A...

Paviland Cave
Kent's Cavern
Arcy-sur-Cure
St.-Césaire
El Castillo
El Pendo
Gato Preto
Cova Beneito
Jebel Irhoud

Trou Magrite
Hohlenstein-Stadel
La Plage
Abric Romaní
Gorham's Cave

Vindija
Riparo
Mochi

Kostienk..
Korolevo
Istállöskö
Pestera cu (...
Temnata   Bacho K..
Cave
Uçagizli Maga..
Ksar Ak..
Skhu..
Qafzel..

120–60,000 YEARS AGO

A F R I C A

Omo Ki..

COASTAL ROUTE TO AMERICA
During the last ice age, North America was dominated by two huge ice sheets. While an ice-free corridor eventually opened up, it is possible that humans may have been able to make use of an earlier route, traveling around the warmer, ice-free coasts.

ATLANTIC OCEAN

S O U T H   A M E R I C A

Pedra Furada

MONTE VERDE
One part of the Monte Verde site in southern Chile is securely dated to around 15,000 years ago. This open-air site has excellent archaeological preservation, with remains of plants used for food or medicine, and tentlike structures.

15,000 YEARS AGO

Monte Verde

SOUTH AFRICAN SITES
The Blombos Caves and Klasies River Mouth at the southern tip of Africa have some of the earliest evidence for complex behavior such as art, jewelry, and bone tools, as well as sophisticated hunting practices involving new projectile weapons.

Blombos Caves
Klasies River Mouth

Conflicting models

What is *Homo sapiens*' relationship to the early hominin species that spread beyond Africa? Historically there have been two competing theories. The Regional Continuity theory suggests that *Homo ergaster* in Africa, *Homo erectus* in Asia, and *Homo neanderthalensis* in Europe each developed over time in these regions into local populations of *H. sapiens*. In contrast, the Recent African Origin model proposes that *H. sapiens* evolved only once, in Africa, and spread out from there to replace all other hominin species across the Old World. The earliest *H. sapiens* fossils and evidence of "modern" behavior are found in Africa, and genetic research on living humans both points to a recent common evolution of every member of our species in Africa and demonstrates the spread of our species across the globe. However, genetic evidence also demonstrates that the evolution of *H. sapiens* was not a straightforward, linear process, and that interbreeding occurred between modern humans and local populations of other hominin species—most notably Neanderthals and Denisovans. Some of their genes live on in us.

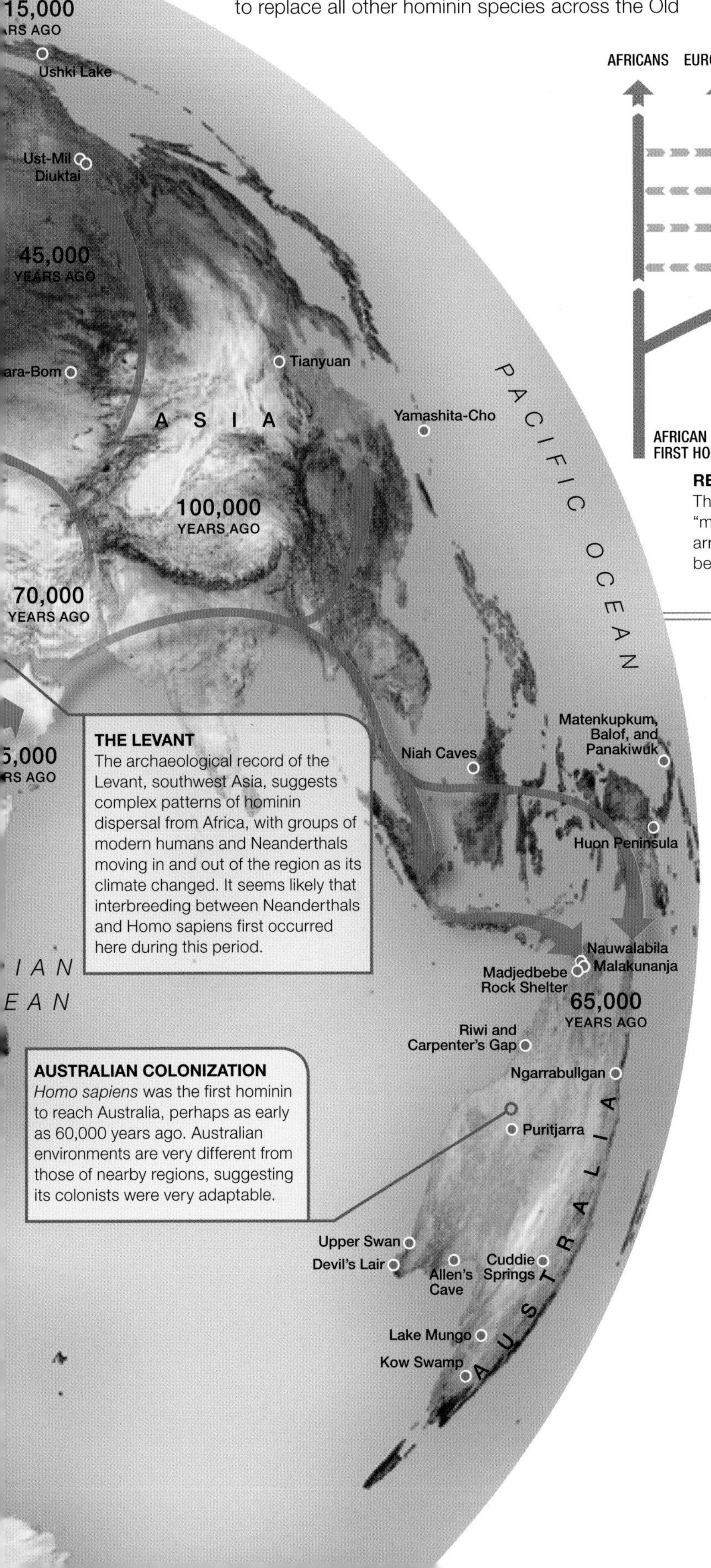

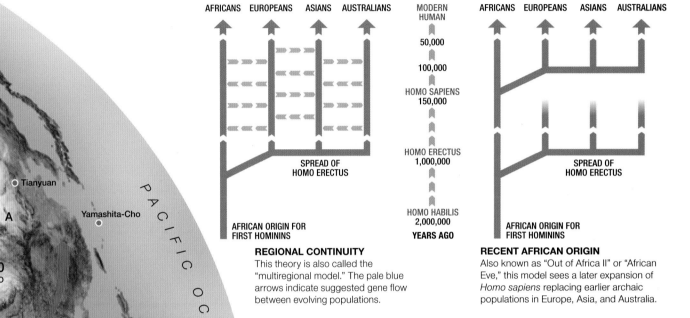

REGIONAL CONTINUITY
This theory is also called the "multiregional model." The pale blue arrows indicate suggested gene flow between evolving populations.

RECENT AFRICAN ORIGIN
Also known as "Out of Africa II" or "African Eve," this model sees a later expansion of *Homo sapiens* replacing earlier archaic populations in Europe, Asia, and Australia.

(Map labels, left page:)

15,000 YEARS AGO
Ushki Lake

Ust-Mil
Diuktai

45,000 YEARS AGO

ara-Bom

Tianyuan

Yamashita-Cho

PACIFIC OCEAN

ASIA

100,000 YEARS AGO

70,000 YEARS AGO

5,000 YEARS AGO

Matenkupkum, Balof, and Panakiwuk

Niah Caves

Huon Peninsula

IAN EAN

THE LEVANT
The archaeological record of the Levant, southwest Asia, suggests complex patterns of hominin dispersal from Africa, with groups of modern humans and Neanderthals moving in and out of the region as its climate changed. It seems likely that interbreeding between Neanderthals and Homo sapiens first occurred here during this period.

Nauwalabila
Madjedbebe Malakunanja
Rock Shelter
65,000 YEARS AGO

Riwi and Carpenter's Gap

Ngarrabullgan

Puritjarra

AUSTRALIAN COLONIZATION
Homo sapiens was the first hominin to reach Australia, perhaps as early as 60,000 years ago. Australian environments are very different from those of nearby regions, suggesting its colonists were very adaptable.

AUSTRALIA

Upper Swan
Devil's Lair
Cuddie
Allen's Springs
Cave

Lake Mungo
Kow Swamp

Reconstructing past migrations

Broadly speaking, the evidence for ancient migrations comes from physical anthropology, archaeology, and genetics; or more simply, from bones, stones, and genes. Anthropologists realized a long time ago that the key to finding out how our species originated and developed lay in looking at the variation between modern human populations.

Historically, physical similarities and differences between groups of people around the world, especially variation in skull shape, have been used to create models of the human family tree. However, recent developments in genetic research have provided a powerful new way of studying human origins and modern diversity. Once again, it is similarities and differences between people that form the evidence, but this time it is the variation in our genes, rather than in our bones, that the researchers are probing.

SECRETS OF HUMAN DIVERSITY

Although humans seem incredibly variable in appearance, more than 99% of our genetic code is identical. This means that there must be small but important differences embedded in our genes. Geneticists are studying those parts of our genomes that vary, and finding the genetic changes that underlie variation in obvious characteristics such as skin color, as well as differences in physiology that, for example, allow some individuals to live comfortably at high altitude.

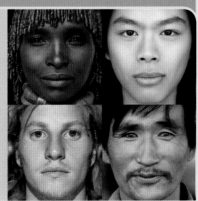

Genetics of past migrations

Studies of the genetic diversity of living people have begun to help tell the story of the origin and spread of our species around the world. In recent years, the contribution of genetics to the pool of evidence has helped achieve some resolution in long-standing debates, although areas of controversy still remain.

Human genetic ancestry

Studies of the genetic diversity of living people—looking at the similarities and differences in DNA from populations all over the world—can be used to reconstruct family trees. There are two general ways of doing this, either by looking at differences in genes between populations and constructing "population trees," or by looking at patterns of mutations in a particular section of DNA, and constructing "gene trees." Gene trees based on small pieces of DNA—in particular, mitochondrial DNA (which we inherit only from our mothers) and part of the Y chromosome (which men inherit only from their fathers)—have proved very useful in studies of human evolution. However, the focus is now shifting to looking at variation across the genome, and a more detailed and complex picture of human genetic ancestry is emerging.

mitochondrion mitochondrial DNA strand

MITOCHONDRIAL DNA
Mitochondria are small capsules inside cells, where sugar is oxidized to release energy for use in the cell. They contain their own DNA—a tiny, circular strand.

human body cell

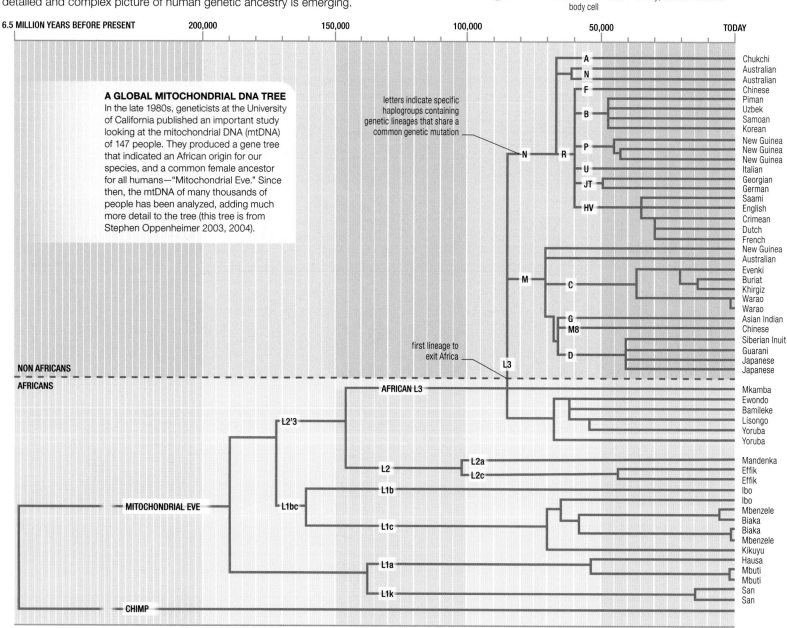

A GLOBAL MITOCHONDRIAL DNA TREE
In the late 1980s, geneticists at the University of California published an important study looking at the mitochondrial DNA (mtDNA) of 147 people. They produced a gene tree that indicated an African origin for our species, and a common female ancestor for all humans—"Mitochondrial Eve." Since then, the mtDNA of many thousands of people has been analyzed, adding much more detail to the tree (this tree is from Stephen Oppenheimer 2003, 2004).

letters indicate specific haplogroups containing genetic lineages that share a common genetic mutation

6.5 MILLION YEARS BEFORE PRESENT | 200,000 | 150,000 | 100,000 | 50,000 | TODAY

NON AFRICANS

AFRICANS

first lineage to exit Africa

INDIVIDUAL TRIBE OR COUNTRY

A — Chukchi
N — Australian
F — Australian
— Chinese
— Piman
B — Uzbek
— Samoan
— Korean
P — New Guinea
— New Guinea
— New Guinea
U — Italian
JT — Georgian
— German
— Saami
HV — English
— Crimean
— Dutch
— French
— New Guinea
— Australian
— Evenki
C — Buriat
— Khirgiz
— Warao
— Warao
G — Asian Indian
M8 — Chinese
— Siberian Inuit
D — Guarani
— Japanese
— Japanese

AFRICAN L3 — Mkamba
— Ewondo
— Bamileke
— Lisongo
— Yoruba
— Yoruba

L2'3
L2 — L2a — Mandenka
L2c — Effik
— Effik
L1b — Ibo
L1bc — Ibo
— Mbenzele
L1c — Biaka
— Biaka
— Mbenzele
MITOCHONDRIAL EVE — Kikuyu
L1a — Hausa
— Mbuti
— Mbuti
L1k — San
— San

CHIMP

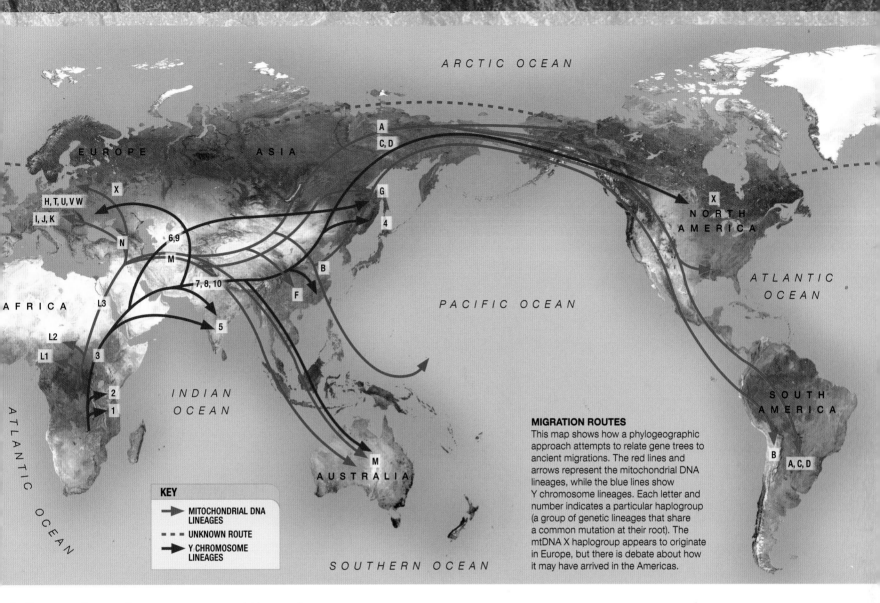

ARCTIC OCEAN

EUROPE   ASIA

| A |
| C, D |

X
| H, T, U, V W |
| I, J, K |

X

NORTH
AMERICA

G

4

N
6,9

M

7, 8, 10

ATLANTIC
OCEAN

AFRICA

L3

B

F

PACIFIC OCEAN

5

L2

3

L1

INDIAN
OCEAN

2
1

SOUTH
AMERICA

B
A, C, D

M

AUSTRALIA

MIGRATION ROUTES
This map shows how a phylogeographic approach attempts to relate gene trees to ancient migrations. The red lines and arrows represent the mitochondrial DNA lineages, while the blue lines show Y chromosome lineages. Each letter and number indicates a particular haplogroup (a group of genetic lineages that share a common mutation at their root). The mtDNA X haplogroup appears to originate in Europe, but there is debate about how it may have arrived in the Americas.

KEY
→ MITOCHONDRIAL DNA LINEAGES
- - - UNKNOWN ROUTE
▶ Y CHROMOSOME LINEAGES

SOUTHERN OCEAN

Mapping human migrations

Genetic diversity has been shown to be greatest among African populations. Because greater diversity indicates an older lineage, this strongly suggests that our species originated on this continent. As well as helping us identify where our species originated, genetic studies can be used to help map past migrations. The gene trees that are constructed from studies of diversity in mitochondrial and nuclear DNA are abstract family trees (like the one shown opposite). Phylogeography attempts to plant those genetic family trees in a geographical and historical context. Branches of the genetic tree are taken to relate to specific events in population history, such as migration. Mitochondrial and Y chromosome DNA trees have been used in this way, to suggest the pattern and timing of ancient migrations that led to the colonization of the world by our species. For example, on the mitochondrial DNA tree shown opposite, the branch labeled L3 indicates the migration of modern humans out of Africa, around 85,000 years ago.

Some geneticists have urged caution with this approach, saying that gene trees are too random to be used to map migrations. The picture will no doubt become clearer as methods improve and larger DNA samples are analyzed. As the focus of research shifts from small regions of DNA to genome-wide analysis, it is becoming clear that any model of ancient colonization must include mixing as well as splitting of populations.

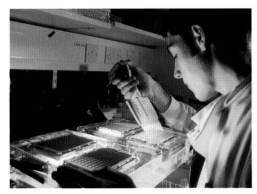

THE HUMAN GENOME PROJECT
This project revealed the complete human genetic code. It was a huge achievement, taking 13 years to complete. Building on the initial map of the genome, geneticists are now trying to understand more about the specific function of genes and other stretches of DNA, and to study genetic variation on a global and genome-wide scale.

GENETICS AND DISEASE

Some of the variation in human DNA is caused by "silent mutations," which have no effect on the way the body works. Other mutations have significant effects and have spread quickly in various populations: they have been selected because they confer an evolutionary advantage.

Scientists searching for evidence of selection in our genes have found that many of the strongest signals relate to resistance to infections. Several different adaptations have evolved in tropical areas, providing resistance to malaria, a disease that may have been affecting us for 100,000 years. Other mutations have been found to provide resistance to infections such as tuberculosis, polio, and measles. Sickle cell anemia (below) provides some protection against malaria, and seems to have arisen independently at least four times, in different African populations.

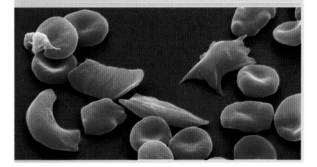

Early hominin migration

The evidence points to the earliest hominin species being exclusively African in their origin and range. However, from around 2 million years ago, we start to see fossil and archaeological evidence of hominins existing outside of Africa, in both Europe and as Asia.

Expansions into Eurasia

At Dmanisi, Georgia, archaeologists have found thousands of fossils and stone artifacts dated to as early as 1.7 million years ago. Prior to this find, it was assumed that *Homo ergaster* was the first hominin species to emerge out of Africa. Some of the Dmanisi skulls do resemble *H. ergaster*, but one looks more like the earlier *Homo habilis*. These fossils may represent separate expansions of each species from Africa. Some researchers even think that australopithecines may have migrated out of Africa.

Two sites in Spain's Sierra de Atapuerca have yielded fossils that show hominins in western Europe by 1.2 million–800,000 years ago. These fossils, named *Homo antecessor*, may represent a European offshoot of *H. ergaster*. Recent discoveries of stone tools at Happisburgh, England, suggest that *H. antecessor* may have reached northern Europe by 1 million years ago. This implies that the species had the ability to make clothes or use fire—essential for keeping warm in northern Europe.

Back in East Africa, there are a few fossil finds that indicate *H. ergaster* persisted in its original homeland until at least 1 million years ago.

KEY

- **HOMO ERGASTER FINDS**
- **HOMO GEORGICUS FINDS**
- **HOMO ERECTUS FINDS**
- **HOMO ANTECESSOR FINDS**
- 💀 **FOSSILS**
- 🝆 **TOOLS**

A MUDDLE OF TOOLKITS

DMANISI CHOPPER

crude form compared to African handax

form resembles African handax

UBEIDIYA HANDAX

AFRICAN HANDAX

If *Homo ergaster* was the earliest African species to colonize the world, we might expect to find similar tools to those used by *H. ergaster* in Africa—Acheulean handaxes—around the globe. Acheulean handaxes have been found at Ubeidiya, Israel, dating to between 1.4–1 million years ago. But at Dmanisi and in East Asia, the earliest stone tools are much cruder, flaked pebbles. Perhaps the toolmakers' ancestors left Africa before handaxes were invented, or people adapted the tools they made to local conditions. East Asian stone tools are crude, but it is possible that hominins there used other materials, such as bamboo, to make tools that have not survived.

ATLANTIC OCEAN

Happisburgh

1,200,000 YEARS AGO

Atapuerca

EUROPE

1,700,000 YEARS AGO

Dmanisi

Ubeidiya

2,000,000 YEARS AGO

2,000,000 YEARS AGO

AFRICA

Buia

Dak

Olorgesailie

Olduvai

HAPPISBURGH
Evidence from this coastal site in Norfolk, England, indicates that ancient humans were living in northern Europe much earlier than previously thought—1 million–800,000 years ago.

DMANISI
At the turn of the 21st century, archaeologists found hominin fossils and stone tools at this medieval fortress site in Georgia. Some experts suggest that the skeletons are of *Homo erectus*, others that they represent a new species: *Homo georgicus*.

OLORGESAILIE
This site in Kenya includes a dense scattering of handaxes. Recent redating of the stone tools places them at 600,000–900,000 years old.

Early humans in East Asia

The earliest human fossils from East Asia come from sites on the Indonesian island of Java, dating to 1.8–1 million years ago. Fossils of *Homo erectus* have also been found in China, many from the Zhoukoudian cave site near Beijing. These "Peking Man" fossils date from 800,000–400,000 years ago.

Some Chinese paleoanthropologists endorse the multiregional model of human evolution (see p.177). They believe that the fossil cranium of Peking Man bears some important similarities to the skulls of modern Chinese people. They see these similarities as evidence that *H. erectus* in China evolved locally into *Homo sapiens*. However, other studies of the fossils and of the DNA of living people, point toward a later influx of modern humans replacing earlier species, rather than regional continuity in East Asia.

ZHOUKOUDIAN

The site of Dragon Bone Hill at Zhoukoudian, China, has produced evidence not of dragons, but of East Asian *Homo erectus*. Hundreds of fossils were discovered there during the 1930s, but most of the originals were lost during World War II.

PEKING MAN

Nihewan ● Zhoukoudian

Lantian

1,600,000 YEARS AGO

A S I A

I N D I A N O C E A N

P A C I F I C   O C E A N

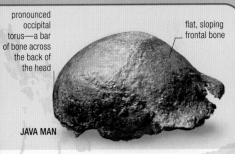

pronounced occipital torus—a bar of bone across the back of the head

flat, sloping frontal bone

JAVA MAN

TRINIL

The Dutch paleoanthropologist Eugène Dubois found a fossil skullcap near the village of Trinil, Java, in 1891. He named it *Pithecanthropus erectus* ("erect ape-man"), but we now know it as *Homo erectus*. Dating the fossil has proved problematic, with estimates ranging from 1.7 million years old to less than 1 million years old.

1,800,000 YEARS AGO

SAVANNAHSTAN

Early fossils of *Homo* from outside Africa have prompted suggestions that we need to be more open-minded about the origin and spread of ancient species. Other groups of animals migrated between Africa and Asia during the Pliocene, and grasslands stretched from west Africa to north China, forming a huge "savannahstan." Although no australopithecine fossils have been found outside Africa, there is no reason to assume that we know the full range of hominin species in Africa and Asia. And if australopithecines made it to Asia, there is even a possibility that *Homo* arose there, and not in Africa.

Sangiran ● ● ● ● Mojokerto

Trinil

SANGIRAN

In the 1930s, the German paleontologist Gustav Heinrich Ralph von Koenigswald found a skullcap at Sangiran, Java, that was almost identical to the Trinil skullcap. When von Koenigswald met Franz Weidenreich, a German anthropologist and anatomist working on the Peking Man material, the pair agreed that both fossils belonged to the same species, now called *Homo erectus*.

A U S T R A L I A

POSSIBLE EXPANSION ROUTES

This map shows some of the sites that have provided evidence for the expansion of early humans out of Africa. The arrows represent hypothetical routes; the actual pattern of dispersal is still hotly debated.

The last of the ancients

Today, modern humans, *Homo sapiens*, are the only surviving hominin species, found on every continent. However, over the last half a million years multiple distinct lineages have co-existed. At various times these included *Homo erectus*, the Denisovans, and *Homo floresiensis* across Eurasia; *Homo heidelbergensis*, evolving into *Homo neanderthalensis*, in Europe; and *Homo naledi* and *Homo sapiens* in Africa.

KEY SITES
This map shows the location of *Homo heidelbergensis* archaeological and fossil sites in Africa and Europe, possible *H. heidelbergensis* sites in Asia, and possible late-surviving *Homo erectus* in Asia.

KEY
- *Homo heidelbergensis* finds
- Possible *Homo heidelbergensis* finds
- *Homo erectus* finds

Swanscombe
Schoningen
Mauer
Vertes-szollos
Arago
Petralona
Atapuerca
Rabat
Zuttiyeh
Jinniushan
Xujiayao
Dali
Yunxian
Maba
Hathnora
Bodo
Omo
Ndutu
Ngangdong
Ngawi
Sambungmacan
Kabwe
INDIAN OCEAN
Elandsfontein

Common ancestor

Around 600,000 years ago, the people of Europe changed significantly both anatomically and behaviorally. Brain size increased, and technology became more refined, with carefully shaped wooden tools alongside the ubiquitous handaxes. Hunting techniques became more accomplished. *Homo heidelbergensis* had arrived. Fossils from the Sima de los Huesos (pit of the bones), in the limestone hill of Sierra de Atapuerca in northern Spain, date from around 430,000 years ago and bear features—such as a projecting face and a bump on the back of the head—that reveal these people to be the ancestors of the Neanderthals.

At the same time, in Africa, *Homo heidelbergensis* is represented by a range of fossil skulls dating from between 700,000 and 400,000 years ago, from various sites including Bodo in Ethiopia and Kabwe in Zambia. As in Europe, the archaeological record reflects an advancement in cognitive abilities. Although fossils from this period in Africa are limited and quite diverse, as a whole they demonstrate a clear pattern of evolution toward modern humans.

HUNTING AND COOKING

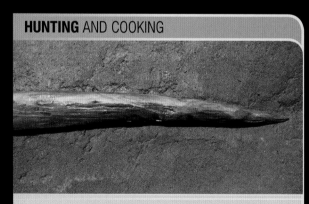

Various discoveries have revealed that, by around half a million years ago, hominins were capable hunters and used fire in a controlled way.

The 400,000-year-old Schoningen spears (seen here) are important archaeological finds. The spears are well balanced for throwing and they show that hominins at this time were making specialized weapons from wood. These people were not just scavengers—they were actively hunting animals.

Sites with possible evidence of controlled fire date back to 1.5 million years, but by 500,000 years ago, there is widely accepted evidence for the use of fire. Cooking is likely to have been important, because it increases the energy to be gained from food.

ENTERING THE SIMA DE LOS HUESOS
This site is a small chamber, which is accessed through a 43ft (13m) vertical shaft. The archaeologists working here are proficient climbers and cavers.

Cousins in the East

The fossil and archaeological record of East Asia during this period is patchy and poorly dated. Nevertheless, there is evidence that *Homo erectus* persisted in Java—from sites at Ngangdong, Sambungmacan, and Ngawi. Various dating techniques have suggested these fossils might date to somewhere between 300,000 and 30,000 years ago.

Several fossil finds from sites such as Dali, Maba, Jinniushan, and Xujiayo, in China, have been put forward as evidence for the multiregional model of human evolution (see p.177). It has been suggested that they represent transitional forms between *Homo erectus* and *Homo sapiens*. The dating of these skulls has proved difficult, but they appear to be between 260,000 and 80,000 years old. These fossil skulls are sometimes claimed to combine anatomical traits of *Homo erectus* and *Homo sapiens*, which might demonstrate local evolution of modern humans in East Asia. However, careful analysis has revealed that these skulls belong either to one species or the other, and that the most convincing explanation for the fossil, archaeological, and genetic evidence from East Asia is that the local hominin populations died out and were replaced by modern humans from elsewhere.

DALI SKULL
This 200,000-year-old skull from China has been put forward as evidence of regional continuity. It seems to retain some archaic features, but with a large braincase.

SIMA DE LOS HUESOS
This cave contains a large collection of bones from many different individuals. It may be a very early example of deliberate disposal of dead bodies, although it is hard to know whether this was motivated by hygiene or ritual.

A new species appears

By 350,000 years ago, the archaic population of *Homo heidelbergensis* in Europe and Africa had evolved in slightly different directions to produce two distinct species. Neanderthals emerged in Europe, while our own species, *Homo sapiens*, evolved in Africa.

MODERN HUMAN BEHAVIOR

These shell beads, which were discovered in Blombos Cave, South Africa, during excavations led by the archaeologist Christopher Henshilwood in 2004, have been dated to around 75,000 years ago. The perforated shells may have been strung together into a necklace. They suggest a modern way of thinking and behaving: they are evidence of ornamentation, and of ritual associated with burial.

African homeland

Fossil, genetic, and archaeological evidence all point toward Africa as the homeland of our species. Although all living humans have a very similar genetic make-up, the genomes (the complete set of genetic material) of African populations are the oldest and most variable, suggesting a longer period of evolution. Transitional fossils documenting the evolution of *H. sapiens* from African *H. heidelbergensis* are known only in Africa, and the earliest *H. sapiens* fossils are also found here. At Jebel Irhoud, Morocco, a very modern-looking skull dates to 300,000 years ago, while similar skulls from Omo Kibish and Herto in Ethiopia are 195,000–155,000 years old. Many of the fossils from this time period show variable mixtures of archaic and modern traits, and the recent dating of *Homo naledi* to this time period has further demonstrated the complexity of the processes by which modern humans evolved.

KLASIES RIVER MOUTH
This site in South Africa includes several caves with deep layers of archaeology, producing evidence of intermittent human occupation 125,000–60,000 years ago. The caves have yielded good evidence for the use of fire, composite weapons, hunting, and shellfish collection.

KEY SITES
This map shows sites where early *Homo sapiens* fossils and transitional fossils have been found, as well as sites with archaeology that has been interpreted as evidence for modern human behavior.

KEY
- Transitional fossils
- Modern human fossils and archaeological evidence for "modern" human behavior

Map labels: Dar-es-soltan, Temara, Jebel Irhoud, Skhul, Hauah Fteah, Qafzeh, Taramsa, Herto, Omo Kibish, Mumba, Laetoli, ATLANTIC OCEAN, Border Cave, Cave of Hearths, Blombos Cave, Florisbad, Die Kelders, Klasies River Mouth, Pinnacle Point

Expanding horizons

A successful species, *Homo sapiens* flourished and eventually pushed out of Africa. The first fossil evidence for any modern humans outside Africa comes from the Middle East, from the sites of Skhul and Qafzeh in Israel, dating to around 120,000 years ago. However, the early evidence of modern humans in the Middle East seems to have been from an expansion that was not maintained. It seems that a switch to a colder climate may have driven those pioneers back to Africa. The expansion of our species out of Africa that eventually led to the colonization of the globe would start later—after 100,000 years ago.

HEINRICH EVENTS

These climatological events involve huge icebergs breaking off North Atlantic ice sheets. These lower the temperature of the sea and produce a switch to a colder, drier climate. Such an event occurred around 90,000 years ago, and the subsequent climate change could have driven our pioneering ancestors back out of the Middle East.

MOUNT CARMEL CAVES
These caves, near the Israeli coast, have been investigated since the 1920s. While Skhul and Qafzeh Caves have produced evidence of modern humans, nearby Tabun Cave contained Neanderthal remains.

BLOMBOS CAVE
This cave site contains a remarkable record of life in ancient South Africa, sealed for thousands of years by a thick layer of wind-blown sand. The hunter-gatherers who used this cave had a varied diet including land animals, fish, and shellfish. They also made what is the first known art: a piece of ocher bearing a scratched, geometric pattern (see p.33).

JWALAPURAM

The super-eruption of the Toba volcano (see below) deposited an ash layer 6 in (15 cm) thick over the entire region of South Asia. Archaeologists digging at the site called Jwalapuram, India, have found tools that they claim are likely to have been made by *Homo sapiens* both above and below the Toba ash layer. Although fossil finds are needed to confirm that the tools are indeed those of modern humans, this discovery could be the oldest evidence yet of modern humans' presence in South Asia.

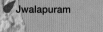

Black Sea

A S I A

Ksar Akil

Boker Tachtit

Red Sea

Jebel Faya

Jwalapuram

INDIAN OCEAN

Lenggong Valley

JEBEL FAYA

A recent archaeological find of stone tools from Jebel Faya rock shelter, United Arab Emirates, has been dated to 125,000 years ago. The tools are similar to those linked with *Homo sapiens* in Africa at about the same time. This raises the possibility that modern humans may have expanded out of Africa much earlier than previously thought.

A F R I C A

Batadomba-lena

BAB-EL-MANDEB STRAIT

While some paleoanthropologists think that the migration out of Africa would have followed a northern route, across the Sinai peninsula, some geneticists have argued for a southern route, across this narrow strait between the Horn of Africa and the Arabian peninsula.

LAKE TOBA

Around 74,000 years ago, a huge volcano erupted on Sumatra, leaving a 62-mile (100-km) wide crater that is now filled by Lake Toba. It was the biggest eruption the human species has ever witnessed. At the time, there were various hominin species living in Eurasia, including the Neanderthals in Europe, the "Hobbit" (*Homo floresiensis*) in Indonesia, and *Homo heidelbergensis* in China.

East along the coast

Some time after 80,000 years ago, a population expansion and migration began, which would lay the foundations for modern humans colonizing the globe. Although the archaeological evidence is not conclusive, the genetic trail leads out of Africa, through the Middle East, into southern and South Asia, and all the way to Australia.

NIAH CAVE
This cave in Borneo was excavated in the 1950s and 1960s by polymath Tom Harrisson, who had received no formal training in archaeology. Harrison discovered the "Deep Skull," which was radiocarbon dated to 40,000 years ago. This finding met with much disbelief at the time, but recent reexcavation and redating have proved Harrisson right.

KEY

→ MIGRATION OF *HOMO SAPIENS*
• SITE OF EARLY *HOMO SAPIENS*
☻ FOSSILS
💧 TOOLS

Niah Cave

Jerimalai

Malakunanja and
Nauwalabila rock shelters

Madjedbebe
Rock Shelter

PACIFIC OCEAN

AUSTRALIA

Mungo and
Willandra Lakes

MUNGO MAN
This skeleton was found in the eroding dunes around the dry Lake Mungo, part of the Willandra Lakes system. Initial dating of the skeleton and the sediment in which it was buried suggested it might be 60,000 years old, but more recent dating has placed it at around 40,000 years old. Together with the cremated remains of "Mungo Lady," these are still the most ancient human remains in Australia.

Arabia—the gateway to the rest of the world

The evidence indicates that modern humans emerged out of northeast Africa into the Middle East, but there is debate as to whether the main route was through the Sinai peninsula, to the north of the Red Sea, or across the southern end of the Red Sea, into Arabia.

Although much of Arabia would have been dry, uninhabitable desert for most of the Pleistocene, recent archaeological evidence suggests that some areas may have remained as green oases, even during arid periods. Furthermore, the lower sea levels of the time would have made the southern coast of Arabia one long oasis—a conduit between Africa and Asia. But there is a problem: while there are some archaeological sites in Arabia, the stone tools they contain could have been made by earlier human species, as well as by *Homo sapiens*. As yet, there is no definite archaeological evidence of modern humans spreading out of Africa, into Arabia, and beyond, but many researchers argue that the genetic data supports just such a model of model human population expansion and migration.

MALAKUNANJA ROCK SHELTER
This site in Arnhem Land has produced artifacts such as stone tools and evidence for pigment use, including grindstones and ocher. The oldest artifacts are dated to 60,000 years ago, and similar dates have been produced for the nearby rock shelter site of Nauwalabila. While some question these dates, it is possible that Australia was colonized by modern humans before Europe.

Farther afield

Geneticists argue that analysis of DNA of living humans indicates that the migration of *Homo sapiens* through Asia probably occurred between 80,000 and 40,000 years ago. Hominins used to living off coastal resources may have spread quite rapidly along the coast of the Indian Ocean. However, archaeological evidence is still very thin on the ground.

Tools found at Jwalapuram, in India, dating to more than 70,000 years ago may indicate a modern human presence, but diagnostic human remains have yet to be found. The oldest firm evidence of modern humans in Southeast Asia is the "Deep Skull" from Niah Cave, Borneo, dating to around 40,000 years ago. Similar dates have been obtained for sites in New Guinea and Tasmania, and for human remains from the Willandra Lakes region of Australia, although archaeological sites in northern Australia have been dated to 65,000 years ago.

The colonization of Europe

Although Europe is very close geographically to Africa, it wasn't colonized by modern humans until after 50,000 years ago. After the first firm, fossil evidence of modern humans in the caves of Skhul and Qafzeh, in Israel, around 90,000–120,000 years ago, there is a gap of tens of thousands of years before the next evidence of modern humans in the Middle East, and then in Europe.

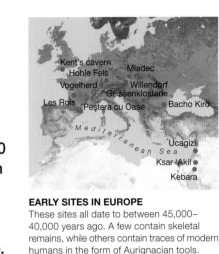

EARLY SITES IN EUROPE
These sites all date to between 45,000–40,000 years ago. A few contain skeletal remains, while others contain traces of modern humans in the form of Aurignacian tools.

The coastal routes and up the Danube

Around 50,000 years ago, an improvement in the global climate, and with it the appearance of habitable landscapes where once there was desert, may have provided the opportunity for modern humans to spread into Europe. Whether these people were spreading north from Arabia, or from northeast Africa, is still debated, but at around 50,000–45,000 years ago, we start to see more evidence of modern humans in the Eastern Mediterranean. Following these glimpses of the colonizers, there are early sites in Turkey, and then a series of archaeological sites across Europe, dating to 45,000–40,000 years ago.

The early modern human sites of Europe suggest that the colonizers moved into the continent along coasts and rivers, as they had done elsewhere. There is a scattering of sites along or close to the Mediterranean coast, in Italy, France, and Spain, along the shores of the Black Sea, and along the course of the Danube. There are only a few sites with skeletal evidence of modern humans, but plenty with archaeological evidence of Aurignacian culture—including the earliest known figurative carvings and musical

THE DANUBE
A chain of important sites close to the Danube, including Oase in Romania, Willendorf in Austria, and Vogelherd in Germany, reveal the importance of this route into the heart of Europe.

instruments—which is thought to have been brought into Europe by modern humans. In most places, there is a sharp break between the technologies associated with Neanderthals, and the arrival of the Aurignacian.

Archaeologists have remarked upon the similarity of the spreading of these early modern human sites and the much later dispersal of Neolithic (early agricultural) populations across Europe—both following routes along the Mediterranean coast and up the Danube, reinforcing the importance of geography in human migrations.

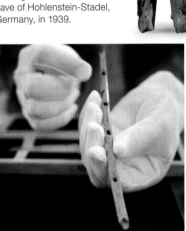

MYSTICAL CARVING
This 30,000-year-old lion-man carving was discovered in the cave of Hohlenstein-Stadel, Germany, in 1939.

EUROPE'S EARLIEST MODERN HUMANS

In 2002, cave divers exploring a cave at Peştera cu Oase, in the southwestern Carpathian Mountains of Romania, discovered a human jawbone. Radiocarbon dating revealed that the jawbone was 35,000 years old. The following year, a team of archaeologists investigated the cave and found pieces of a human skull, which proved to be even older—around 40,000 years old. The skull and mandible are the earliest modern human fossils in Europe, but they have some anatomical quirks, which a few anthropologists have interpreted as evidence of interbreeding with Neanderthals (see also pp.190–91).

MAKING MUSIC
In 2004, archaeologists found ivory and bone flutes at Vogelherd Cave, Germany, dated to around 35,000 years ago—the earliest evidence we have for music-making.

Neanderthals and modern humans in Europe

When modern humans made their way into Europe about 50,000 years ago, they were entering the territory of another human species. Neanderthals and their ancestors had been Europe's indigenous people for hundreds of thousands of years. What happened when the two species met is the subject of intense debate.

"suprainiac fossa"—an indent at the back of the head

rounded chin

Meeting the locals

Recent radiocarbon dating of modern human and Neanderthal sites in Europe suggests that there was an overlap of about 6,000 years, when both species existed together in this continent. The old idea that Neanderthals may have evolved into Cro-Magnons (modern humans) in Europe has long been overturned, but debate rages over the nature of the replacement of the Neanderthals by the incoming modern humans. For some time, there appeared to be a very clear archaeological divide between Neanderthals and modern humans, but the picture has become more complicated in recent decades. The archaeology—from sites such as St-Césaire, Grotte du Renne, and Roc-de-Combe—suggests that Neanderthal culture and technology may have been quite similar to that of modern humans at the time.

KEY SITES
Human remains from Lagar Velho, Peştera cu Oase, Cioclovina, and Mladec may be Neanderthal/modern human hybrids; the archaeological sites of Châtelperron, Roc-de-Combe, and St-Césaire contain Châtelperronian tools (see opposite).

Hybrids

Some Paleolithic modern human bones from Europe show features more usually associated with Neanderthals. Skulls from Mladec in the Czech Republic have an "occipital bun" (a bulge at the back of the head), remains from Peştera cu Oase in Romania exhibit extremely large third molars (wisdom teeth) and a broad upper part of the jaw, while a child's skeleton from Lagar Velho in Portugal also displays some Neanderthal-like traits. Some anthropologists have long argued that these individuals are hybrids, or their descendants, showing that Neanderthals and modern humans interbred, while others have suggested that archaic features could simply have resulted from shared ancestry between our closely-related species. However, geneticists have found that up to 4 percent of modern Europeans' DNA comes from interbreeding with Neanderthals, reopening the debate about these possible hybrids. And in some cases, ancient DNA extracted from fossils has helped to solve the puzzle: nearly 10 percent of the genetic make-up of the Oase individual is Neanderthal in origin.

ATLANTIC OCEAN

Mladec
Grotte du Renne
Sesselfelsgrotte
Châtelperron
Cioclovina
Vindija
St-Césaire
Roc-de-Combe
Peştera cu Oase
Abri Dubalen
Caune de Belvis
Lagar Velho
Mediterranean Sea
Cavallo
Klisoura
Lakonis

KEY
- Sites with transitional toolkits
- Sites with possible Neanderthal/modern human hybrids

short limbs in proportion to the rest of the body

occipital bun

MLADEC SKULL
This skull from a site in the Czech Republic appears to be that of a modern human, but it also bears some apparently Neanderthal characteristics, such as an "occipital bun"—a bulge at the back of the head.

LAGAR VELHO CHILD'S SKELETON
This skeleton, discovered at the base of a cliff in Portugal, is dated to around 30,000 years ago. Some nthropologists argue that the limb proportions and indent on its skull are Neanderthal-like, indicating interbreeding between modern humans and Neanderthals.

Genetic investigation

Recent advances in DNA techniques have enabled scientists to extract genetic material from fossil bones, including Neanderthals and the hominins from Sima de los Huesos (see p.153). Comparing this with modern human DNA shows that the modern human and Neanderthal lineages diverged more than 500,000 years ago, and between 1 and 4 percent of the DNA of modern Europeans and Asians comes from Neanderthals. This indicates recent interbreeding between our species, which may even represent assimilation of the Neanderthals into the modern human population in Europe and Asia. Interbreeding probably happened soon after the modern human population expanded out of Africa, around 50,000–65,000 years ago. The DNA of modern Africans contains traces of other interbreeding events, too, with as yet unknown archaic human populations.

DNA EXTRACTION
Ancient DNA—if it is present at all in a fossil bone—is highly fragmented. The genome (about 3 billion base pairs) must then be reconstructed from tiny fragments of perhaps a hundred base pairs each.

THE "DENISOVANS"—A NEW SPECIES?

Extraction of DNA from a 41,000-year-old finger bone and two teeth from Denisova Cave in Siberia has led to the discovery of a completely new species, known as "Denisovans." Traces found in other ancient and modern DNA sequences from Spain to East Asia suggest that it was quite widespread. Although anthropologists still do not know what they looked like, Denisovans were closely related to, and interbred with, Neanderthals. Both species probably evolved from a common ancestor around 430,000 years ago: those that spread East evolved into Denisovans, while those in the West became Neanderthals. Denisovans also interbred with *Homo sapiens*, leaving traces of DNA in the genomes of modern humans from Southeast Asia, including Melanesians, and Aboriginal Australians. Their DNA also contains evidence of a further, unknown source, suggesting that there may be more undiscovered species of hominins waiting to be found.

Neanderthal toolkits

Historically, archaeologists have considered the toolkits of Neanderthals (see p.154) to have been vastly inferior to those produced by modern humans. When the toolkit called the "Châtelperronian" was first discovered in France, it was assumed to have been made by modern humans. However, later finds and improved dating techniques have led many to conclude that it is actually a Neanderthal toolkit. Along with the Uluzzian culture of Italy and Greece, the Châtelperronian appears to show the Neanderthals as having been more ingenious and creative than previously thought. Interestingly, these toolkits and cultures only appear after modern humans have entered Europe, and some experts have argued that the Neanderthals must have had contact with modern humans, and learned how to make these tools from them. Others still call into question the association of Châtelperronian toolkits with the Neanderthals.

CHATELPERRONIAN FIND
The discovery of Châtelperronian tools close to a Neanderthal burial at La Roche-à-Pierrot (St-Césaire) led archaeologists to conclude that Neanderthals made these tools. But the archaeology involved complex layering, and some experts do not accept the association.

STONE TOOLS

ARTIFACTS
As well as tools, the Châtelperronian culture includes personal ornaments, such as pendants made from animal teeth. It seems that their makers were just as interested in image and personal adornment as we are.

ANIMAL-TOOTH PENDANTS

North and East Asia

The mitochondrial DNA lineages (see pp.178–79) of North and East Asia suggest that these areas were populated by descendants of the original colonizers who spread along the coast of southern Asia. These people may have followed rivers inland, taking them northward around and through the Himalayas, and westward from the east coast of Asia.

KEY SITES
This map shows important sites with Upper Paleolithic archaeology (indicative of modern humans) dated to between 40,000 and 20,000 years ago. A few sites, such as Tianyuan, Yamashita-cho, Malaia Syia, Zhoukoudian, Maloialomanskaia, and Mal'ta, also contain human remains. The Upper Paleolithic first appeared around Lake Baikal 45,000–40,000 thousand years ago, and in eastern sites at later dates.

Moving north

Siberia—a very different environment compared with the tropical coast of southern Asia—presented several significant challenges to the colonizers. They would have had to adapt their culture and technology in order to survive in a place where plants for food and wood for building and fuel were often scarce, and where it could be bitterly cold and incredibly dry. And yet the dates of archaeological finds in Siberia indicate that humans had reached southern Siberia more than 40,000 years ago, and were existing right up in the far north, on the coast of the Arctic Ocean, by 30,000 years ago.

BEADS

IVORY PLAQUE

MAMMOTH-BONE HUTS
Ancient Siberians used mammoths for food and warmth—and also as building materials, as shown by the mammoth-bone huts at sites like Mezhirich. Some mammoths may have been hunted, but it is possible that the Siberians were also scavenging what they needed from frozen carcasses, or collecting bones and tusks washed up in rivers.

SIBERIAN TREASURES
These ivory beads and plaque are from Mal'ta, near Irkutsk in Siberia. The site preserved the remains of a Paleolithic camp, with semisubterranean houses, thousands of stone tools, and over 500 artifacts made of bone, ivory, and antler. Mal'ta dates to about 21,000 years ago—right at the peak of last Ice Age.

People living in the Arctic, such as the Evenki of northern Asia, may have evolved specific adaptations to help them cope with the cold. Geneticists have discovered mutations in mitochondrial DNA that could make the mitochondria generate heat, helping to keep the whole body warm.

MAMMOTH
These Siberian giants had mostly died out by around 10,000 years ago.

Heading east

Analyses of skull shape and genetic data suggest that modern East Asians are not the direct descendants of *Homo erectus* in China—they are more recent colonizers, who can be traced back to the Out Of Africa expansion between 80,000 and 50,000 years ago. The earliest fossil remains of modern humans in East Asia date to around 40,000 years ago.

The archaeology of East Asia is very puzzling; there is no sudden change in stone tools with the arrival of modern humans, who indeed seem to have been making very crude pebble tools right up until around 30,000 years ago. It is possible, however, that the early colonizers were making more complex bamboo tools, which have not survived in the archaeological record.

LIUJIANG SKULL
This modern human skull is controversially dated to 100,000 years ago. Bones from Tianyuan and Laibin, in China, have been more reliably dated to 40,000 years ago.

The New World

Apart from Antarctica, North and South America were the last continents to be colonized by *Homo sapiens*. As far as we know, no other human species made it to the Americas. There is still much debate about the arrival and spread of humans in the New World, although, as with other areas, the addition of genetic clues has shed some light on this particular ancestral landscape.

Routes into the Americas

For much of the Pleistocene (2.6 million–12,000 years ago), Asia and North America were linked by another huge continent—Beringia. It would have been possible for colonizers to make their way overland from northeast Asia into what it is now Alaska. However, during glacial periods, large ice sheets would have blocked the route into the rest of the Americas.

By about 14,000–13,500 years ago, the ice had melted sufficiently to open an ice-free corridor across northwest North America. However, there are now well-dated archaeological sites predating the opening of this route by at least a thousand years. Environmental analyses on the Haida Gwai islands reveal that the ice was retreating from the west coast of Canada 17,000–14,500 years ago. This means that colonizers could have entered the rest of North America by boat along the Pacific coast using frequent landings. Another suggested entry route is from Europe, across the North Atlantic. This route has been proposed due to the similarity between the Clovis (in North America) and Solutrean (in Europe) stone points. However, many experts are sceptical about this link, in the absence of any hard evidence.

EARLY AMERICAN FOSSIL
This skull was discovered in the Sumidouro Cave in Brazil, and dates to about 13,000 years ago. "Luzia," named after the famous "Lucy" skeleton, is among the oldest American human fossils known.

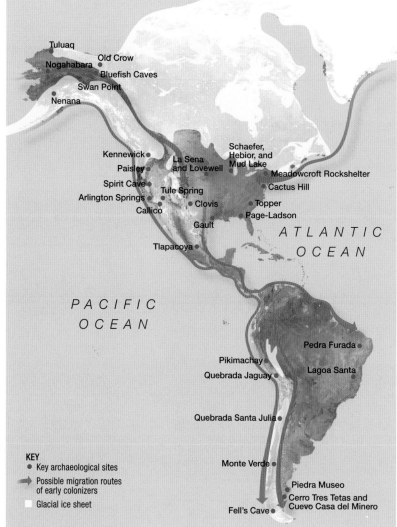

MIGRATION ROUTES
This map shows the great ice sheets that would have blocked the way into the Americas during glaciations, as well as the ice-free corridor, and important archaeological sites. The evidence supports a north-to-south expansion of modern humans into these continents, and the arrows indicate hypothetical routes of colonization.

KEY
- Key archaeological sites
- Possible migration routes of early colonizers
- Glacial ice sheet

TOPPER SOUTH CAROLINA

The Topper site contains Clovis stone tools, as well as pre-Clovis tools dating to about 15,000 years ago. But in 2004, archaeologist Albert Goodyear announced the discovery of much earlier archaeology at Topper, dating to an astounding 50,000 years ago. Other experts have cast doubt on this claim, challenging both the dating and the nature of the archaeology itself.

NATURAL OR MAN-MADE?
The ancient stone "tools" found at Topper may be natural stone flakes rather than human-made tools.

BEAUTIFUL SPEARHEADS
For decades, archaeologists believed that the makers of Clovis points were the first inhabitants of the Americas, about 13,500 years ago. But more recently, with the discovery of older archaeological sites, the "Clovis first" theory has been dismissed.

The first Americans

Genetic evidence supports a model of the colonization of the Americas from northeast Asia. There are a few sites in North America that have been suggested to date to before the Last Glacial Maximum (LGM), 40,000–20,000 years ago. There are problems with dating some sites, and concerns that some "tools" were naturally produced, but a date of 24,000 years ago is widely accepted for the Bluefish Caves site in northern Yukon, Canada.

The earliest, widely accepted sites, dating to 15,000–14,000 years ago, include Meadowcroft Rockshelter and Cactus Hill in the US, and Monte Verde in Chile. Monte Verde is remarkable for the excellent preservation of its archaeology, including hide coverings of huts, medicinal plants, and the earliest evidence of humans using potatoes.

The earliest human remains in the Americas—our first glimpse of the people themselves—include a skull nicknamed "Luzia" from Brazil, dating to about 13,000 years ago (see opposite); the "Anzick baby" skeleton from a rockshelter in Montana, dating to 12,600 years ago; and "Kennewick Man" from Washington State, which is around 9,000 years old. DNA extracted from the Anzick Baby and from Kennewick Man has been found to be similar to that of living Native Americans.

Broadly, the genetic and archaeological evidence is consistent with an initial colonization of northwest North America by at least 24,000 years ago, and an expansion southward, into the rest of North America and South America, from around 16,000 years ago.

ARCH LAKE WOMAN
Evidence of burial ritual appears in the form of these talc beads and tool, which were found in the grave of a woman buried around 10,000 years ago in New Mexico. Her body was also covered with red ochre.

AMERICAN EXTINCTIONS

During the Ice Age, the Americas were home to many species of huge animals including saber-tooth cats, mammoths, mastodons, and the giant sloth. But by around 13,000 years ago, all those animals had disappeared. Some experts have suggested that climate change—perhaps even linked to an asteroid explosion—may have wiped out these animals. But a more likely explanation might involve the arrival on the scene of formidable human hunters.

UNALASKA EXCAVATIONS
Archaeological excavations on the Aleutian Islands have uncovered evidence of a human presence on the coast going back 9,000 years. At least 25 prehistoric villages have been discovered close to what is now Dutch Harbor airport, Unalaska.

Oceania

Many of the islands of Southeast Asia, including Borneo, Sumatra, and Java, were joined to the continental landmass of Sunda during much of the Pleistocene. Some of the islands of Oceania were also joined to larger landmasses: New Guinea and Australia were part of the continent of Sahul. Other Pacific islands, to the east, were always separated from the larger landmasses.

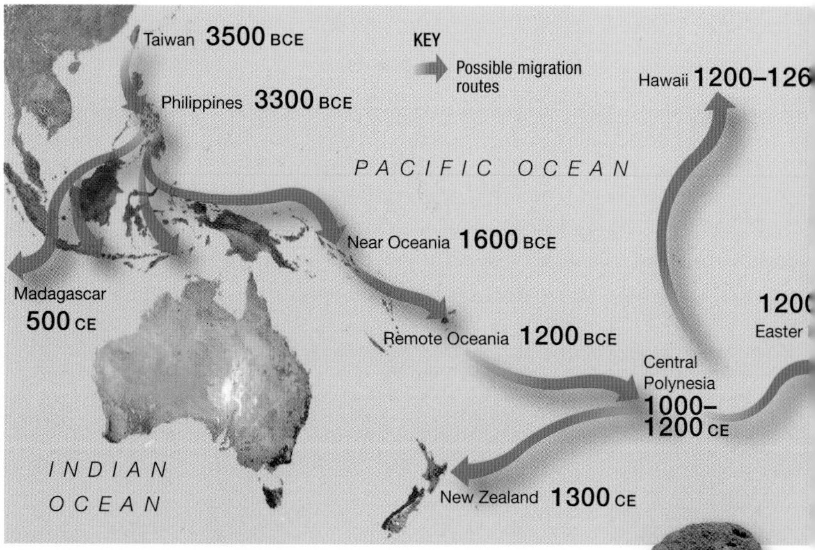

Colonization of Near Oceania

There is archaeological and genetic evidence for early colonization of islands that would either have formed part of Sunda or Sahul, or lay between these continents (see pp.186–87). From the continent of Sahul, Pleistocene hunter-gatherers were able to colonize the islands of Near Oceania, reaching the Bismarck archipelago, off the northeast coast of New Guinea, by 33,000 years ago. The Papuan languages of New Guinea and Island Melanesia seem to be echoes of this ancient phase of colonization.

The early modern human inhabitants of Island Southeast Asia and Near Oceania were hunter-gatherers who lived in coastal and forest environments, making simple pebble and flake tools—or, in some areas, managing to get by with no stone tools at all. However, it seems that the

people who had colonized these islands found it difficult to survive in those unstable coastal environments during the peak of the Last Glacial Maximum (LGM; around 21,000–18,000 years ago). Signs of human presence in New Guinea and the Bismarck Archipelago tend to date to either much earlier or later than the LGM. Between 14,000–7,000 years ago, coastal environments were stabilizing, and shell middens in New Guinea, Timor, and Thailand reveal that modern humans were making good use of the resources to be found in the expanding estuaries and lagoons.

ROUTES THROUGH OCEANIA

This map of the South Pacific Ocean shows the possible migration routes taken by early colonizers through the islands of Oceania. The routes and dates are estimates based on archaeological, genetic, and linguistic evidence.

Map labels:
Taiwan **3500** BCE
Philippines **3300** BCE
KEY
→ Possible migration routes
Hawaii **1200–126**
PACIFIC OCEAN
Near Oceania **1600** BCE
Madagascar **500** CE
Remote Oceania **1200** BCE
120
Easter
Central Polynesia **1000–1200** CE
INDIAN OCEAN
New Zealand **1300** CE

TREE TOPPLER

Recent discoveries have shown that, almost 50,000 years ago, people were living in the highlands of New Guinea, using stone axes like this.

LAPITA CULTURE

Some experts have suggested that this culture, dating from 1350–750 BCE, was carried by the wave of Neolithic colonizers spreading east from the islands of Southeast Asia. However, others have suggested that it originated within the islands, and that its distribution closely follows the colonization pattern revealed by mitochondrial DNA lineages, from Near to Remote Oceania. The Lapita culture includes characteristic pottery, such as the sherds shown here, with stamped decoration.

LAPITA POTTERY

Remote islands

The more remote Pacific islands were reached only very recently. The first evidence of humans living in New Zealand, Easter Island, and Hawaii dates to within the last 1,500 years. The archaeological, linguistic, and genetic evidence seems to suggest that, from 6,000 years ago, Neolithic farmers spread from southern China and Taiwan to Polynesia, apparently without interbreeding with local Melanesian peoples, bringing with them Austronesian languages.

HAWAIIAN PETROGLYPHS
Rock carvings have been found on many Hawaiian Islands. Radiocarbon dating suggests that these stick figures were created in about 1400–1500 CE.

Language and genes

There has been much debate over how quickly colonization happened, and how much the suggested second wave of Neolithic colonizers mixed with the earlier populations in Near Oceania. There have always been discrepancies between the cultural (archaeological and linguistic) and biological (including genetic) evidence, but this is not surprising. Genes are inherited, whereas culture and language can spread between genetically different populations. The Austronesian languages of Southeast Asia and Oceania seem to emanate from the region of Taiwan, and support a model of rapid Neolithic colonization. However, the genetic lineages of many of the people of Oceania are traceable back to the ancient, Pleistocene colonization. The spread of farming and Austronesian languages was more a movement of ideas and culture, than of the people themselves.

CONTACT WITH SOUTH AMERICA

SWEET POTATO

Although archaeological, genetic, and linguistic evidence practically rules out any major South American expansion into Oceania, the journey to Easter Island would have been conceivable. Indeed, it is very likely that there was some early contact between South America and the islands of Oceania. There are striking similarities between ancient Peruvian stone carvings and those found on Easter Island, for instance. Archaeologists have also found very early evidence of sweet potatoes and bottle gourds in Oceania, which must have arrived there from South America.

EASTER ISLAND STATUES
These statues are among the most iconic symbols of human colonization and fragility on Earth. Erected around 1000–1650 AD, it seems that the society that produced this monumental art had already collapsed by the time of European contact in 1722.

FROM HUNTERS TO FARMERS

Technological innovation has been key to the success of modern humans as a species, enabling people to do things beyond their unaided physical capabilities. In the most recent 10,000 years of the human story, the rate of change has increased exponentially, and with it human populations. Agriculture developed in many areas soon after the Ice Age; city life emerged 5,000 years later. And human evolution is far from over—our species has continued to adapt to cope with changes in diets,

MELTING GLACIER

As glaciers melted, huge quantities of water were released, raising global sea levels and expanding or creating rivers, marshes, and lakes. The water also escaped into the atmosphere, returning to the Earth in the form of rain.

SEA LEVEL CHANGES

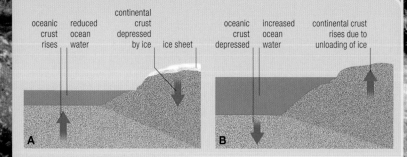

A — oceanic crust rises | reduced ocean water | continental crust depressed by ice | ice sheet

B — oceanic crust depressed | increased ocean water | continental crust rises due to unloading of ice

During an ice age (A) the volume of ocean water is low because water is locked up in ice sheets. When the ice melts (B), the oceans expand, raising global sea levels and drowning much of the continental shelf. As a result, the gradient of rivers is reduced, making them slower and wider, creating floodplains and deltas. In the areas that had been beneath the ice sheets, the land rises (isostatic recovery); in Scandinavia, this process is still continuing.

After the ice

As global temperatures rose and the ice sheets melted, the world underwent many changes, including rising sea levels, increased rainfall, and the spread of forests and grasslands. These offered new opportunities for humans that had a major impact on their way of life.

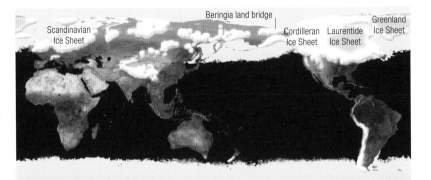

RETREATING ICE SHEETS
When the ice sheets retreated, previously glaciated areas in the north became habitable, but many formerly habitable regions and connecting land bridges were submerged.

KEY
☐ Icecap 18,000 years ago
▨ Icecap 10,000 years ago

Changing landscapes

After the Last Glacial Maximum (LGM), around 16,000 BCE, global temperatures rose, reaching several degrees higher than today by around 7000 BCE. The ice sheets retreated and the water locked up in them returned to circulation, greatly increasing rainfall. Sea levels gradually rose from more than 395 ft (120 m) below present levels at LGM to 165 ft (50 m) below at 9000 BCE, reaching current levels sometime after 4000 BCE. As a result, large areas of coastal land were progressively drowned, particularly in low-lying regions. Beringia, the massive land bridge linking Asia with North America, was submerged around 8500 BCE. Around half the land in Southeast Asia was drowned, creating a mass of islands, and Japan was severed from mainland East Asia. Oceania was equally affected (see p.196). As the ice sheets over

ANTLER HARPOON
In 1931, this harpoon was dredged from the North Sea, which was dry land in 10,000 BCE when it was made.

grooves

barbs

Scandinavia and Scotland melted, the vast plain exposed between them became a low-lying marshland, known as Doggerland, through which a great river flowed. As sea levels rose, this basin was drowned, creating the North Sea. Britain was separated from mainland Europe around 6500 BCE. In Scotland and Scandinavia, however, the rising sea level was offset by isostatic recovery (see left).

DROWNED COAST
The continental shelf was progressively drowned, creating a shallow marine environment ideal for shallow-water fish and shellfish—good food resources for many postglacial hunter-gatherers.

EEL TRAP
Ancient traps similar to this modern eel trap have been found in northern Europe. They were made by people who colonized the region as it became warmer.

willow basketry

trap entrance

Colonizing new lands

Higher temperatures and increased rainfall had a major impact on the environment, vegetation, and fauna. Extensive freshwater lakes were created by glacial meltwaters. Marine resources proliferated in coastal shallows drowned by the rising sea levels. Forests spread from the restricted areas where they had survived the glacial conditions. These changes created new opportunities for human communities, who colonized areas to exploit the new resources on offer. In Europe, the widespread tundra of the glacial period retreated to the far north, a region previously under ice that people now colonized. Animals that had thrived in glacial conditions, such as reindeer, also moved north or became extinct, like the giant elk. Other animals, including red and roe deer, cattle, and boar followed the spreading vegetation. For a few millennia, the Mediterranean was covered by deciduous woodland, rich in edible species, but by 7000–5000 BCE these forests were shifting north into temperate Europe, to be replaced in the Mediterranean by the far less productive evergreen vegetation that is still found there today.

As the Laurentide and Cordilleran ice sheets retreated in the far north, many colonists spread rapidly through the Americas, reaching Tierra del Fuego by 9000 BCE. Clovis hunter-gatherers moving into North America exploited a wide range of plants and animals, including the big game that had lived there during the glacial period (see pp.194–95). In North Africa, increased rainfall and a northward shift of the rain-bearing winds transformed the Sahara Desert, creating lakes, rivers, and grasslands (see p.212).

FRESHWATER LAKES
Melting glaciers created lakes that by 8000 BCE had become rich in aquatic creatures and plants, which attracted wildfowl and other animals. Lake margins were a favored location for foragers.

Hunter-gatherers

Humans adapted their food acquirement strategies and ways of life to exploit the new postglacial environments. While mobility was usual, in favorable areas with exceptional resources communities could become sedentary.

The seasonal round

Hunter-gatherers used a great variety of foods. In most regions, some foods were available in different places at certain times of year. For example, herd animals moved between summer upland and winter lowland pastures, marine fish and wildfowl migrated seasonally, and woodlands produced autumn fruits and spring bulbs. Other food resources, such as shellfish, small mammals, and freshwater fish, could be obtained year-round, as required. Foragers therefore moved in a regular but flexible pattern calculated to make best use of these foodstuffs while also obtaining other useful materials, such as stone for tools.

HONEY GATHERING

Bones and shells in their sites are the most obvious remains of what hunter-gatherers ate, although some traces of plants may survive, particularly nutshells and carbonized seeds. But for much of their diet, the evidence is elusive. This rock painting from Cueva de la Araña in Valencia, Spain, gives a rare glimpse—a figure up a tree gathering honey from a wild bees' nest.

SPRING

Oysters were a critical resource for a few weeks before the arrival of fish and plant foods in May.

SUMMER

Summer coastal resources included cod and mackerel, while many types of fruit were available inland.

WINTER

Winter brought dolphins, porpoises, and gray seals, and migratory birds, such as whooper swans.

AUTUMN

This was the richest season, with harp seals, hazelnuts and acorns, and deer and eels at their best.

SEASONAL RESOURCES IN ERTEBOLLE

In the well-known Danish Ertebølle culture many sites were occupied seasonally, to exploit particular resources. For example, people camped inland at Ringkloster during the winter to catch wild piglets, collect nuts, and trap pine martens for furs.

Forager landscapes

Foragers selected environments offering a diversity of resources, such as lakes, rivers, marshes, coasts, and woodland margins. Particularly favorable were places where these intersected, such as river estuaries, yielding freshwater, land, and marine resources. Dense forests offered little, and were avoided, but there is evidence that postglacial foragers deliberately managed forests to create clearings, encouraging edible plant growth and attracting game.

GARBAGE DUMP

Shell middens—great mounds of discarded mollusk shells, such as these near Glesborg in Denmark—are the most visible remains of postglacial hunter-gatherers' meals, although the shellfish themselves contributed only a small part of the diet.

Settled hunter-gatherers

In some favored regions, with abundant resources that were available year-round or could be stored, foragers could remain in one location. In the eastern Mediterranean, for example, Natufian hunter-gatherers could gather wild cereals, legumes, and nuts when they ripened in sufficient quantities to last the community throughout the year, supplemented by game, fish, and fresh plant foods. In parts of Northern Europe, many hunter-gatherer groups established permanent base camps, usually on the coast, at or within easy reach of places where they could gather shellfish, catch inshore fish and scavenge beached whales, hunt birds, and collect plant foods. They used boats to catch deep-sea fish and marine mammals and hunted large mammals and small game from the adjacent land. From the base camp, small parties could make expeditions to places where other resources could be found. In northwest North America by around 500 CE, a huge annual harvest of salmon swimming upriver to spawn, supplemented by marine and forest resources, supported a prosperous hierarchical society of hunter-gatherers living in a permanent settlement.

ANIMAL TRAPS
Hunter-gatherers in the eastern Mediterranean efficiently hunted gazelle using game drive lanes. The animals were stampeded between converging lines of stones into traps ("kites") like this one, where they were killed.

MOUNTAIN GAZELLE

Bows and arrows became important in the postglacial period, enabling game to be killed at a distance with less effort and greater accuracy than a spear. Hunting large game has probably always been a male activity—and it still is for the San of southern Africa. Except in circumpolar regions, however, most of the forager diet comes from plants, aquatic resources, and small game, usually collected, trapped, or hunted by women and children.

Way of life

Mobile communities only owned things that they could carry with them. These may have included tools, along with elaborate shell, bone, antler, and stone jewelry and decorations sewn onto clothing. These people enjoyed full lives that were rich in social, cultural, and spiritual activities, such as singing, dancing, and storytelling. Their intimate knowledge of their environment was often reflected in complex mythology. Spiritual beliefs and ties with particular places were sometimes expressed in the creation of cemeteries. In Atlantic Europe, for example, people were buried with grave goods for use in the afterlife.

While in some seasons groups might split into small bands to exploit widely scattered resources, during others, when food was plentiful within a small area, several groups might congregate, providing the opportunity for social interaction and also marriages. Kinship ties were important for survival, because they allowed individuals to turn to relatives in other groups for help in time of need.

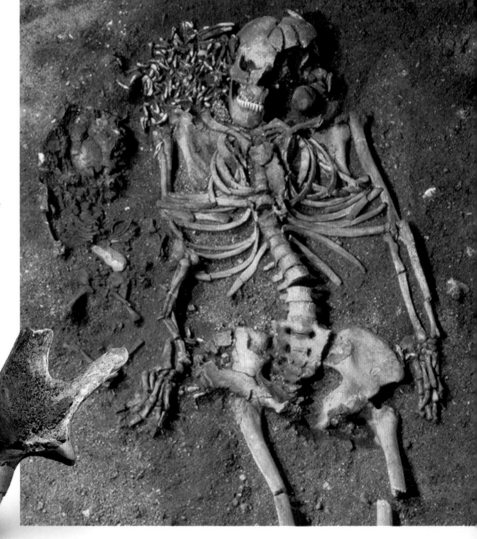

CHERISHED DEAD
In this grave at Vedbaek, Denmark, a newborn boy was laid to rest on a swan's wing, beside his young mother.

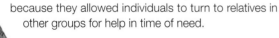

antlers deliberately broken off

RED DEER SKULL-CAP
This headdress, from Star Carr in Yorkshire, England, provides a rare glimpse of ritual activities. It may have been worn by a shaman or in dances related to hunting magic.

holes cut into cranium for a leather thong

Rock art

Rock paintings and engravings provide a vivid picture of the people of the past, engaged in their daily activities and customary rituals. The art also reveals the world as they saw it, including their mythology, so interpreting rock art involves the difficult task of disentangling reality from imagination and interpreting ancient imagery.

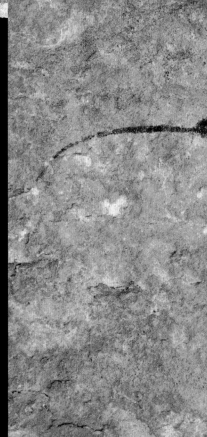

Art has been executed on rock surfaces for tens of millennia. Most early examples come from caves and rock shelters where they are protected from the elements, though some have been found in open-air sites, particularly engravings. Although more durable than paintings, these may be hard to distinguish from surface features created by weathering, so often escape notice. Rock art traditions are particularly rich in certain parts of the world, including prehistoric engravings in Scandinavia and the Italian Alps and paintings and engravings in Mesoamerican caves. Dating the art is difficult, particularly in places such as Australia, central India, and southern Africa where there is an unbroken tradition stretching back to the Palaeolithic. Clues used in dating include the technology and activities shown in scenes, for example, chariots in Saharan paintings indicate an Iron Age or later date; stylistic features that change through time; and associated archaeology, such as the date of deposits covering the art. Rock art provides invaluable clues to the way of life and environment of the communities that created it, for example, the green Sahara region of the early postglacial period. Many activities of which no archaeological trace survives are portrayed in art, such as dancing and singing —Bhimbetkar dancers, for example, have song bubbles by their mouths. Yet the meaning of the art may be very different from its surface significance: seemingly mundane objects and creatures may reflect mythology and rituals that are beyond the grasp of the modern observer.

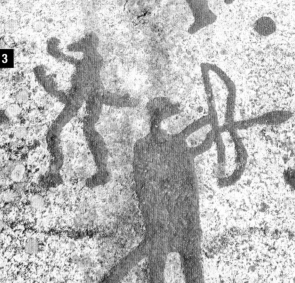

1. Inanke Cave, Zimbabwe. Eland (depicted bottom right in this rock painting) are common in San art. In southern African Bushman traditions, these antelope provide a link with the spirit world.
2. Kagga Kamma rock shelter, South Africa. Paintings of Bushman foragers often depict their daily activities. Here one man carries a stick or bow, the other a game bag.
3. Rocks at Fossum, Sweden. These Bronze Age Scandinavian rock engravings show men, deer, and boats. All probably have deeper meanings: the boats may symbolize the sea or the Sun's daily journey.

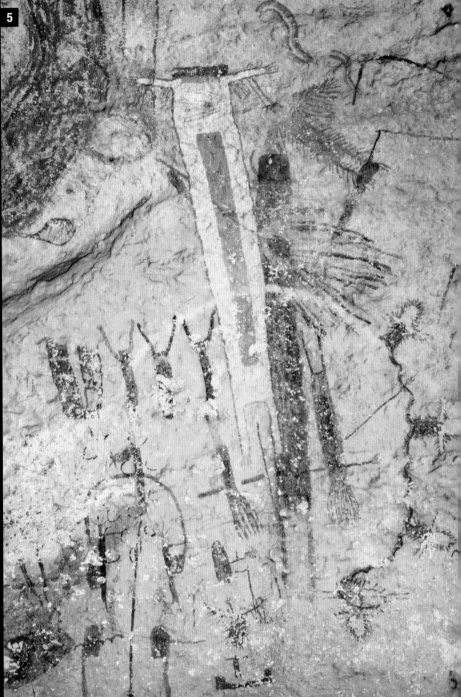

4 5

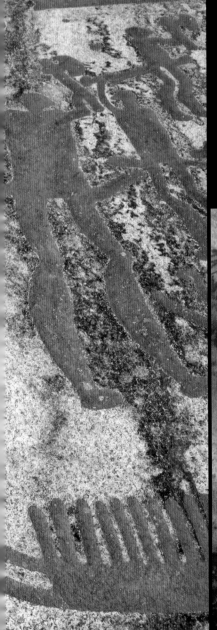

4. Tassili N'Ajjer rocks, Algeria. Saharan art often shows men hunting (as here) and many animals no longer found in the region.
5. White Shaman Cave, Texas, USA. Lower Pecos rock art depicts animals, shamans and their equipment, and activities such as gatherings of people in headdresses.
6. Oenpelli rocks, Arnhem Land, Australia. Plant and mineral pigments were used in ancient Aboriginal rock art such as this painting of a turtle.
7. Bhimbetka rock shelter, India. The medieval art in red shows a procession of warriors on horses; the fainter white paintings of people and animals are prehistoric.

7 6

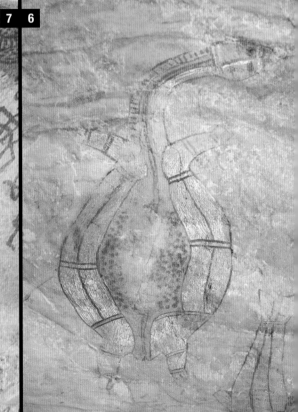

From foraging to food production

Agriculture began independently in many parts of the world, for varying reasons, with different domesticated species, and with different paths of development. It was not a postglacial discovery—hunter-gatherers already knew a great deal about plants and animals and often manipulated them or the environment to increase productivity.

Crop cultivation

Uneven preservation of plant remains has focused attention on grains and pulses, but new techniques allow microscopic remains of tubers and other plants to be detected, producing a more balanced picture of early cultivation. High-yielding cereals, such as wheat, barley, rice, and various millets, became important staples for postglacial foragers in many regions, providing storable food. Planting harvested grain led to genetic changes in the plants, and beneficial changes were encouraged by selective planting. For example, tiny teosinte spikes were transformed over millennia into large maize cobs that provided a staple for Mesoamerican farming (see p.218). Legumes (such as peas and beans), which are rich in protein, were often cultivated alongside cereals. In tropical forest areas, tubers, such as yams, and tree crops, such as sago, were cultivated, but many underwent little genetic change. Over time a huge range of other plants were domesticated for food, medicine, dyes, and other uses.

WILD EINKORN
The stalks of einkorn, an early wheat, broke when the grain ripened, scattering the seed. From mutant plants that failed to break, farmers bred domestic einkorn that stood till harvested.

EINKORN SEEDS

EINKORN
SEED HEADS

Animal husbandry

Only certain animals are suitable for domestication, mainly gregarious species that are amenable to herding. Domestication therefore took place only in some parts of the world. Mesoamerica, for instance, lacked suitable herd animals, and farmers continued to rely on wild sources of animal protein. Separation from the wild breeding population and the selective pressures of human management, including deliberate breeding, brought about the changes seen in domestic animals. Sheep and goats were domesticated in two or possibly three places in the western half of Asia and spread widely from there; pigs, however, were domesticated in many places in the Old World. Llamas and alpacas were the only herd animals domesticated in the New World. Wildfowl and other birds were domesticated in many regions.

MINOAN BULL
LIBATION VESSEL

DOMESTIC ANIMALS
Early domesticated species such as sheep and cattle were selectively bred for smaller size and increased docility. This is particularly marked in the transformation of the huge, fierce wild aurochs into the domestic cow.

EASTERN NORTH AMERICA

Goosefoot
Knotweed
Little barley
Maygrass
Squash
Sumpweed
Sunflower

SOUTHWEST NORTH AMERICA

Agave	Turkey
Amaranth	
Tepary bean	

MESOAMERICA

Avocado
Bottle gourd
Capsicum
Common bean
Corn
Scarlet runner bean
Squash
Tomato

YUCATAN PENINSULA

Cacao
Cotton

SOUTHERN AMAZONIA

Chili peppers
Common bean
Manioc
Peanut
Warty squash

TROPICAL LOWLANDS

Arrowroot
Butternut squash
Chili peppers
Common bean
Guava
Manioc
Sieva bean
Sweet potato
Yam

ANDEAN ZONE

Amaranth	Alpaca
Coca	Guinea pig
Lima bean	Llama
Cotton	Muscovy duck
Jack bean	
Potato	
Quinoa	
Squash	
Tobacco	

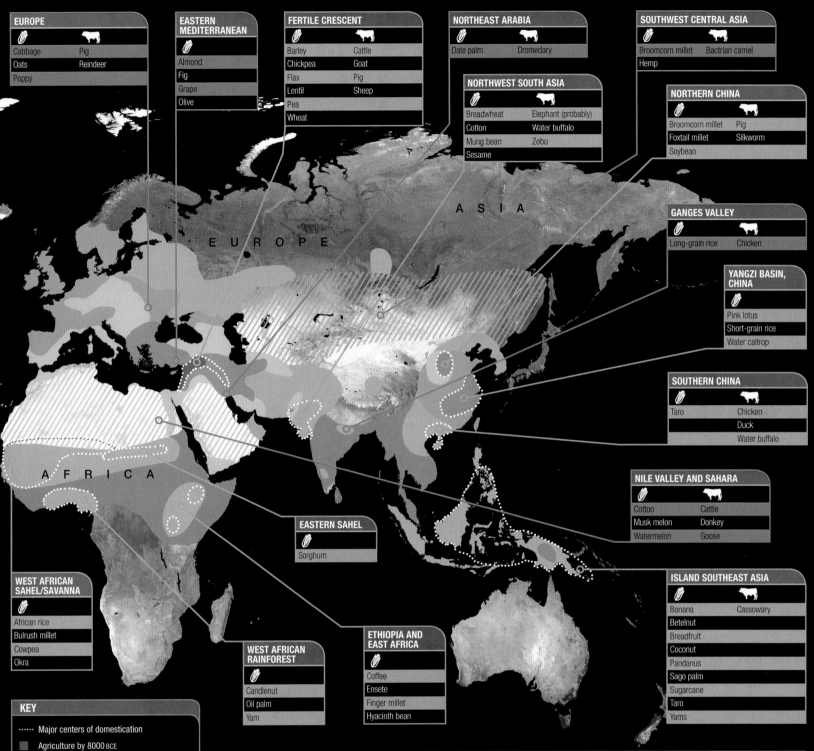

EUROPE

Cabbage	Pig
Oats	Reindeer
Poppy	

EASTERN MEDITERRANEAN

Almond
Fig
Grape
Olive

FERTILE CRESCENT

Barley	Cattle
Chickpea	Goat
Flax	Pig
Lentil	Sheep
Pea	
Wheat	

NORTHEAST ARABIA

Date palm	Dromedary

NORTHWEST SOUTH ASIA

Breadwheat	Elephant (probably)
Cotton	Water buffalo
Mung bean	Zebu
Sesame	

SOUTHWEST CENTRAL ASIA

Broomcorn millet	Bactrian camel
Hemp	

NORTHERN CHINA

Broomcorn millet	Pig
Foxtail millet	Silkworm
Soybean	

GANGES VALLEY

Long-grain rice	Chicken

YANGZI BASIN, CHINA

Pink lotus
Short-grain rice
Water caltrop

SOUTHERN CHINA

Taro	Chicken
	Duck
	Water buffalo

NILE VALLEY AND SAHARA

Cotton	Cattle
Musk melon	Donkey
Watermelon	Goose

EASTERN SAHEL

Sorghum

ISLAND SOUTHEAST ASIA

Banana	Cassowary
Betelnut	
Breadfruit	
Coconut	
Pandanus	
Sago palm	
Sugarcane	
Taro	
Yams	

WEST AFRICAN SAHEL/SAVANNA

African rice
Bulrush millet
Cowpea
Okra

WEST AFRICAN RAINFOREST

Candlenut
Oil palm
Yam

ETHIOPIA AND EAST AFRICA

Coffee
Ensete
Finger millet
Hyacinth bean

KEY

- ······ Major centers of domestication
- Agriculture by 8000 BCE
- Agriculture by 6000 BCE
- Agriculture by 5000 BCE
- Domestic animals only by 5000 BCE
- Agriculture by 4000 BCE
- Agriculture by 2000 BCE
- Pastoralism by 2000 BCE
- Agriculture by 1000 BCE

THE SPREAD OF FARMING

Plants and animals were domesticated in many regions. Some were then spread by farmers colonizing new areas or by trade. Sometimes these replaced less productive local domesticated species. As they spread into different environmental zones, new varieties developed to suit new conditions.

Why farm?

The benefits of farming may seem obvious—higher yields and a reliable food source under human control. However, farming is often harder work and increases risk—foragers exploit a much wider range of resources, so they have alternatives if some fail. Population growth may have put pressure on resources, making farming preferable to infanticide. Farming could enable people to live in one place for longer periods, by providing storable food for when wild resources were scarce, or bringing preferred foodstuffs closer to home. Domestic animals offered a reserve food supply, to be eaten or traded when times were hard. Both crops and animals may have been raised to provide feasts, an important social activity.

SLASH AND BURN

In this millennia-old form of tropical farming, trees are cleared and burned, killing the seeds of weeds and creating ashy soil that is very fertile for three to five years.

Farmers in West and South Asia

In the early postglacial period, farming communities appeared in eastern Turkey and areas to the south, growing emmer and einkorn wheat, barley, flax (for oilseeds and fibers), lentils, chickpeas, beans, and other legumes. By 7000 BCE, farming had spread more widely in West Asia, from central Turkey as far east as southwest Iran, and domestic sheep, goats, pigs and cattle were also kept.

KEY SETTLEMENTS
The earliest farming settlements lie in an arc in West Asia known as the Fertile Crescent. From here, farming spread into other parts of West Asia, through Turkey, Iran, and beyond; it also developed independently in parts of South Asia.

West Asia

The late glacial period saw the emergence of permanent hunter-gatherer villages in the hilly woodlands of West Asia, depending on, among other resources, wild wheat, barley, and legumes. After 9600 BCE, in the aceramic or Pre-Pottery Neolithic, their successors started to cultivate these plants. They still hunted, but also began herding sheep and goats; cattle and pigs were domesticated a little later.

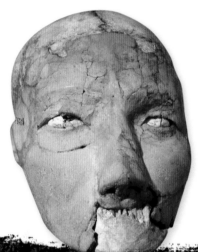

SKULL FROM JERICHO
In Pre-Pottery Neolithic settlements, the dead were often buried under the houses. Some skulls were removed and facial features modeled in clay over them, often with cowrie shells in the eye-sockets.

People mostly lived in small villages of round or rectangular houses, such as Tell Abu Hureyra (Syria), often located near water. However, Jericho (Israel) was much larger at 6 acres (2.5 ha). It was surrounded by a massive stone wall with an impressive tower, possibly protection against flooding. Twined and woven linen textiles were now being made.

After 7000 BCE, farmers expanded into new regions, including the north Mesopotamian plains, and pastoralism became important in some areas. Clay was used to make simple pottery containers; by 6000 BCE these were both technologically sophisticated and attractively decorated with painted designs.

CYPRIOT NEOLITHIC HOUSES
Cyprus was colonized by 9000 BCE. These houses of stone and plastered mudbrick have been recreated in the 7th-millennium BCE village of Khirokitia.

Çatalhöyük

Çatalhöyük is an extraordinarily large settlement for its time. Dated 7400–6200 BCE, it was 32 acres (13 ha) in extent and was occupied by up to 8,000 people, all living in similar households, with no evidence of a social hierarchy. Although favorably situated in the fertile Konya Plain (Turkey), which provided land for growing crops and pasture for many sheep, the settlement's size is still puzzling. The house interiors were plastered and richly decorated with paintings showing possibly religious scenes (such as a huge bull surrounded by puny humans), geometric designs, and figures of mud plaster, including bulls' heads modeled over real skulls with horns. The dead were buried beneath the floor of certain houses.

SPOON

SPATULA

FORK

BONE UTENSILS
Many domestic objects and tools have been found at Çatalhöyük, which is still being excavated. These bone utensils were probably shaped using obsidian tools (see p.226).

SETTLEMENT RUINS
The houses were tightly packed, with adjoining walls and without streets: access was over the flat roofs, from which ladders led down into the house interiors. Each had a main room, with solid benches around the walls, and usually a smaller storeroom.

Central and South Asia

From West Asia, farming spread across the northern Iranian Plateau, reaching the eastern Caspian by 6000 BCE, where sites like Djeitun have the West Asian suite of domestic crops and animals. Only gradually did farming reach other parts of the arid Iranian Plateau. At Mehrgarh in Pakistan, however, a site was found dating back to 7000 BCE with both wheat, barley, legumes, sheep and goats, and locally domesticated cattle. It is still unclear whether the people here developed farming independently, how they acquired wheat—which was certainly introduced—and whether the other crops and animals were also introduced or locally domesticated. Mehrgarh, which was originally occupied seasonally, grew into a substantial village by 5500 BCE, when other farming villages are also known in the region. Elsewhere in South Asia, rice farming began by the 3rd millennium BCE in the Ganges Valley, while in South India foragers who kept cattle also began growing local plants around this time.

Farming had some adverse effects on health. The constant labor of grinding grain caused women to develop arthritis. A diet of cereals was often low in protein and vitamins, causing blood and bone disorders, such as anemia and osteoporosis, and it promoted tooth decay. The first known dental work comes from Neolithic Mehrgarh, where holes were drilled in teeth with a stone drill.

ZEBU CATTLE
Changes in bones at Mehrgarh chart the gradual domestication of the zebu, the humped cattle of South Asia. It became the region's principal domestic animal.

MEHRGARH RUINS
Both mudbrick houses and compartmented storage buildings (shown here) are found at Mehrgarh. It was occupied continuously for more than 4,000 years, eventually growing into a sizable town.

Göbekli Tepe

Recent discoveries have opened a window onto the extraordinary ritual life of early postglacial foragers and farmers in West Asia. At Göbekli Tepe, in Turkey, circular enclosures containing massive, exuberantly decorated T-shaped pillars were constructed more than 11,000 years ago.

Göbekli Tepe was discovered in 1994 by archaeologist Klaus Schmidt, who has been excavating it with a German-Turkish team ever since. He believes it was a central holy place for communities who came together from across a wide area. No settlements have been found nearby. Remarkably, construction began here before 9500 BCE, preceding agriculture in the region. Some unfinished pillars lie where the T-shaped limestone slabs were quarried, close to the site. Constructing the circular buildings and shaping, carving, moving, and erecting the pillars, which were up to 16 ft (5 m) high and 10 tons in weight, would have involved the collaborative efforts of hundreds of people, something archaeologists had not expected of the small communities of this time.

Most of the stones are carved with wild animals and other creatures, and geometric patterns. Schmidt has suggested that the animals may have been the totems of the different groups who combined to build the enclosures, or were typical fauna of the regions from which they came. Human elbows, shoulders, and fingers carved on some of the stones suggest that the pillars themselves were intended to represent humans in symbolic form. Buildings of a later phase, dated around 8000 BCE, are square rather than circular and are much smaller, with pillars up to 5 ft (1.5 m) high and mainly without animal decoration. Eventually, in the 8th millennium BCE at the latest, the enclosures were deliberately filled in with domestic garbage, including many wild animal bones, rocks, and earth.

1. Excavation site. Excavations run for four months each year. Site surveys indicate that there were at least 20 enclosures, of which six have been uncovered.
2. Predator sculpture. This snarling creature has been called a reptile, lion, or leopard. Many creatures here are menacing: perhaps the art was an attempt to dominate dangerous aspects of the artists' world or to engage these as protectors of a sacred site.
3. Limestone T-shaped pillars revealed. Some pillars have shallow cupmarks on their top. They stand in enclosures ranging from 33 ft (10 m) to more than 65 ft (20 m) across.

4. Circular enclosure. A deep circular pit dug into the ground was lined with drystone walls into which T-shaped pillars were set. Two larger pillars were erected in the building's center. Additional walls were later built inside the enclosure, reducing the interior space.

5. Fox pillar. A fox decorates each of the paired pillars in this enclosure's center. Foxes were clearly important in both ritual and daily life. Only snakes are depicted more frequently than foxes, and many fox bones were found in the earth used to fill the enclosures.

6. A complex scene. Boars, cranes, vultures, other birds, scorpions, and snakes, all depicted on this stone, are common motifs, as are geometric patterns, though the design on each stone is different.

6 5

KEY SETTLEMENTS
Agriculture developed in the north of Africa and spread south. Farmers and herders gradually replaced a rich mosaic of well-adapted forager communities.

Farmers in Africa

Farming began in North Africa by 5000 BCE, using both local and West Asian crops and animals. The presence of East and West African crops in India around 2000 BCE indicates that farming must have begun in those regions before then, although only later remains have been found. Farming gradually spread, reaching southern Africa in the early 1st millennium CE.

The Saharan Lakes

Increased rainfall after the Ice Age created grasslands, lakes, rivers, and marshes in what is now the Sahara Desert. It was populated by animals that are now confined to sub-Saharan regions, including hippopotamuses and elephants. People here and in the adjacent Nile Valley began making pottery, decorated with wavy lines, as early as 9000 BCE, probably for cooking plant foods. The lakes were rich in fish, such as Nile perch, which they caught in large numbers as well as gathering plants and hunting game—a lifestyle they depicted in their rock paintings. By around 7000 BCE, they also began herding local cattle, perhaps to enable them to cope with the short-term fluctuations in climate that frequently occurred.

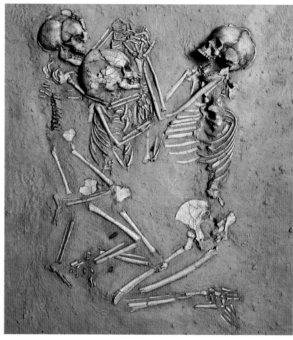

PEOPLE OF THE GREEN SAHARA
Foragers buried in a large cemetery at Gobero before 6000 BCE had lived by a deep lake. By 4000 BCE, the lake was shallow, with small fish, and the people living here also kept cattle. In this poignant burial a woman and two young children hold hands.

AERIAL VIEW
Traces of the Sahara Desert's green past are revealed by satellite photography. These 10 small lakes in the Ounianga Basin in Chad are the remnants of a huge lake that existed here around 12,800–3500 BCE.

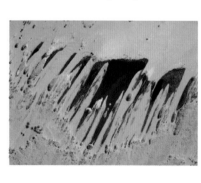

POSSESSIONS AND PROPERTY

Living year round in one place allowed people to accumulate possessions, particularly objects, such as this quern, that were too heavy for mobile communities to transport. The ability to store property encouraged ideas of ownership and inheritance; these grew as agriculture became more intensive and people invested labor in long-term improvements to pieces of land.

grindstone

quern

North Africa

As the Sahara became increasingly arid after 6500 BCE, the grasslands retreated southward and shrank in extent and cattle pastoralism became more important. Sheep and goats, introduced from West Asia around 5000 BCE, were also herded by various communities across North Africa. Other introductions included wheat, barley, flax, and pulses and these became staples of Nile Valley agriculture. Once agriculture was firmly established in Egypt and Nubia, local animals and plants were also domesticated, including geese and varieties of melons. An important local domesticated species, by 3500 BCE, was the donkey, which then spread throughout West Asia. Farming settlements prospered, particularly in Egypt where the annual flooding of the Nile River provided both water and fertile silt for raising crops; the river was also a rich source of fish and wildfowl.

Sub-Saharan Africa

Cultivation in West and East Africa probably began around 3000 BCE. Each area had different domesticated crops, including pearl millet in West Africa, sorghum in eastern Sudan, ensete in Ethiopia, and finger millet in East Africa. Pastoralism—raising cattle, sheep, and goats—was important throughout the region. Farther south, in the rain forest, yams and tree crops such as oilnut palm were probably cultivated from a similar date. In the Niger Delta, African rice was domesticated during the 1st millennium BCE. Banana phytoliths (microscopic silica bodies) recently discovered at Nkang (Cameroon), dated around 500 BCE, are evidence of early long-distance contacts: bananas are native to Southeast Asia.

FOOD STAPLE
Sorghum is a staple throughout Africa and in parts of Asia. It can be made into hot cereal, couscous, bread, or beer.

HERDING CATTLE
Many rock paintings in the Sahara, like this example from Tassili Maghidet in Libya, show people with their cattle, driving herds to pasture, or engaged in their daily activities in camp while many cattle wander among them.

TRADITIONAL GRANARY
A major advantage of agriculture is that some crops, such as cereals, can be stored for use long after harvest. Storage generally involves keeping the crops dry, cool, and free of pests, in covered pits, storage vessels, or granaries. Traditional African granaries, such as this one in Burkina Faso, are built of braided reeds or grasses, with a conical grass roof.

POPULATION GROWTH

Since mobile groups must carry children who are too young to walk, they tend to space births at intervals of two or more years. Settled communities, however, have no such constraints, so women may bear a child each year. Living in one place, therefore, was widely associated with population growth. In addition, agriculture involved more work than foraging, and children could usefully contribute to this, for example by herding the domestic animals, so there was an incentive for farmers to produce large families.

Farmers in East Asia

Cereal cultivation began early in China and spread into Korea, Japan, and mainland Southeast Asia—regions with well-developed postglacial forager communities. Rice, pigs, and chickens also spread into Island Southeast Asia, where tubers and tree crops had been cultivated for thousands of years. Pottery was widespread from early times.

EXCAVATED HOUSES AT BANPO
Forty-six houses, most of them round but some square, were excavated in the moated village of Banpo in the 1950s. Some had their floors partially below ground; others were built at ground level.

Yellow River Valley

The inhabitants of the Yellow River (Huang He) basin in northern China cultivated broomcorn millet possibly before 8000 BCE and by 6000 BCE were also growing foxtail millet, various fruits and vegetables, and keeping pigs. Substantial villages, such as Banpo (see left), appeared by 5000 BCE. The villagers made silk and hemp textiles and attractive red and brown pottery, fired in kilns outside the village. Adults were buried in an external cemetery while children were interred in large urns within the village. Differences in grave goods reflected an emerging social hierarchy. Agriculture was the mainstay of the economy, though many wild resources were still used. Pigs were the main animals, but chickens, water buffalo, and a few cattle were also kept. Rice spread into northern China by 3000 BCE.

KEY SETTLEMENTS
From China, farmers spread south and west. Foragers in regions to the northeast took up farming, using Chinese crops and animals, which also spread to the northwest. Farmers and fishers who colonized the Pacific took with them mainland animals and plants of Island Southeast Asia.

Yangzi Valley

Short-grain (*japonica*) rice was cultivated in the central and lower Yangzi Valley by or before 6000 BCE. A vivid picture of life is provided by the waterlogged site of Hemudu, where ropes and many wooden objects have been preserved, including spades, oars, and the remains of timber longhouses. Domestic pigs were raised, but wild plants, game, fish, and marine mammals were also important. By 4000 BCE stone-tipped wooden plows were being used to prepare ground, presumably drawn by domestic water buffalo, although evidence for these is known only later. Foxtail millet was introduced into the middle Yangzi by 4500 BCE but did not spread to the lower Yangzi. Jade jewelry and fine pottery found in some but not all graves indicate a growing social hierarchy.

BONE TOOLS
Animal shoulderblades from Hemudu were securely fastened to wooden handles for use as spades or hoes. Spades were also made of wood or stone.

STILT HOUSE
The houses in Yangzi villages were raised on piles to keep them above the level of the river's seasonal floods, as can be seen in this reconstruction at the Hemudu site museum.

Southeast Asia

Rice farmers from the Yangzi region spread along river valleys into southern China and Vietnam by about 3500 BCE and southwest into Thailand and Cambodia, where rice cultivation was gradually adopted after 2300 BCE by local communities, such as Khok Phanom Di, who exploited wild coastal and forest resources. By 3000 BCE, farmers growing foxtail millet and rice and keeping pigs and chickens spread into Taiwan, reaching the Philippines and Island Southeast Asia by 2000 BCE. Here, a separate farming tradition, based on yams, taro, and tree crops such as sago and coconut, was already well established—horticulture in the New Guinea highlands dates to 7000 BCE. These plants, along with dogs, pigs, and chickens, were the staples of the farmer-fisher groups who spread out into the Pacific after 1500 BCE.

RED JUNGLE FOWL
Chickens were domesticated from red jungle fowl separately in southern China, by 6000 BCE, and in India before 2500 BCE.

BURIAL AT KHOK PHANOM DI
This woman was buried with fine pottery and shell jewelry, including large ear ornaments and 100,000 beads, probably sewn onto her clothes.

Korea and Japan

The postglacial foragers of Korea and Jomon Japan enjoyed a productive lifestyle, hunting, gathering plants such as nuts, utilizing marine and aquatic resources, and often living in permanent coastal or riverside communities. The Jomon made elaborate pottery, which they used for cooking shellfish and plant foods. They also cultivated some local plants by the late 4th millennium BCE. At Sannai Maruyama, a settlement of 50–100 households, they built a 48 ft (14.7 m) high structure of six massive wooden posts, perhaps a watchtower or shrine. Millet spread into Korea by 4000 BCE and later Japan. During the 3rd millennium BCE, rice spread into Korea, where it rapidly became important. Wheat, barley, and cannabis—introduced to China from the west via Central Asia—also reached Korea at this time, as did bronze metallurgy. Rice, wheat, and metallurgy spread to Japan in the early 1st millennium BCE.

COOKING AND STORAGE

Mobile people need portable containers, like bags and baskets. But sedentary communities can use heavier materials, such as stone, and more fragile ones, such as clay. Pottery (fired clay) first appeared in Japan around 14,000 BCE; the pot shown here is a later example of the distinctive Jomon pottery. Useful for general storage, pottery is also watertight and fire resistant, so it can be used to heat liquids. Pottery probably brought about a culinary revolution: boiling made food more digestible and let ingredients be stewed together.

Farmers in Europe

Farming entered Europe around 7000 BCE and was the main way of life across Europe by 4000 BCE, although wild resources were still important in many regions. Farmers with West Asian domestic plants and animals colonized some regions, while in others local foragers adopted these domesticated species. Native European species were also domesticated.

KEY SETTLEMENTS
As farming settlements spread across Europe, they adapted to the challenges of the different climates and environments.

HAMBLEDON HILL
Neolithic causewayed enclosures in Britain served various purposes. Here, they included a settlement and an area where the dead were exposed before selected bones were buried in the long barrow visible toward the rear of this later, Iron Age hillfort.

Greece and the Balkans

Farmers from Anatolia (Turkey) spread over Greece and the Balkans between 7000 and 6000 BCE, settling on fertile plains, where they grew wheat, barley, and legumes, and raised sheep and goats. They lived in villages of rectangular mudbrick or clay-daubed timber houses. Long-lived settlements, particularly in the Balkans, such as Karanovo, grew into substantial mounds (tells). The people made plain pottery and, later, geometrically painted pottery and figurines depicting people and animals (see p.222). Some figurines may have been connected with religion; for example, a number were found in a large building at Nea Nikomedeia, which has been interpreted as a shrine. Tools were made of wood or stone, including obsidian (volcanic glass) from the island of Melos—this may have been obtained through exchange networks that linked farmers and Mediterranean hunter-gatherers, although coastal farmers were also competent seafarers.

HOUSE INTERIOR
The single-roomed houses contained a hearth or clay ovens and pottery for cooking and storage, as in this example from northern Greece.

The Mediterranean

Mediterranean foragers came increasingly to rely on fish after 7000 BCE, particularly tuna, a deep-sea species. Seafaring promoted interregional contacts and the exchange of goods and ideas. Pottery making was one of these, and a distinctive type, Impressed Ware, began to be made by Mediterranean communities. Gradually, West Asian animals and crops were also spread, at first supplementing wild resources, but increasing in importance through time. Some farmers from Anatolia or southeast Europe may also have colonized parts of the Mediterranean, particularly southern Italy, where by 6000 BCE they established settlements enclosed by ditches. Farming spread slowly into northern Italy and the interior of Spain and France.

DECORATING TOOL
The edge of a cockle shell was the most common of various sharp tools used to impress designs on early Mediterranean pottery.

CHAMBER TOMB
Later farmers built many types of megalithic tomb, often containing communal burials, in the western Mediterranean (like this Corsican dolmen) but mainly in Atlantic Europe.

Central, Western, and Northern Europe

Farming communities spreading into Central Europe around 5600 BCE had to adapt to bitter winters, heavy rainfall, and dense forests. They kept mainly cattle and farmed open river terraces, where they built small linear settlements of rectangular longhouses with steeply pitched roofs and overhanging eaves, using wood for many purposes. Known as Linearbandkeramik (LBK) from their line-decorated pottery, these people spread rapidly, reaching the Paris Basin by 5200 BCE. Farming spread through Western Europe and into other parts of Central, Northern, and Eastern Europe by 4000 BCE as farmers colonized new areas and hunter-gatherers shifted to a mixed or agricultural economy. Polished stone axes were in common use for forest clearance and woodworking; the stone for these was quarried and widely traded.

DOMESTICATING PIGS
European farmers raised West Asian crops and animals but also cultivated some local plants and, in some regions, domesticated wild boar.

DOMESTIC ARCHITECTURE

Foragers can sometimes settle in one place but this is usual for farmers, who store grain and other produce for year-round use. Settled communities could invest labor and resources in building houses with permanent fittings. Furniture is rarely preserved, but one exception is at Skara Brae in Scotland, where it was made of stone. The houses have dressers, box beds, shelves set in the walls and water-tight boxes in the floor, seats, and a central hearth.

Wetland settlements

Exceptional preservation through waterlogging of marsh and lakeside settlements give a detailed picture of early European life. They demonstrate the continuing importance to European farmers of wild plants and waterfowl, game, and fish. Linen cloth and many different wooden tools are among the objects found in these settlements.

In the Somerset Levels of western England, the timbers of wooden trackways crossing areas of marsh show that Neolithic people here practiced sophisticated woodland management, including coppicing.

UNDERWATER REMAINS
Preserved underwater in Lake Mondsee, Austria, these piles would once have supported Late Neolithic houses on the lake's shores, raising them above the high-water level.

Farmers in the Americas

In significant contrast to the Old World, domesticable herd animals were largely absent from the Americas. Except in the Andes, farmers obtained meat by hunting, often using dogs, which were later used as draft and pack animals.

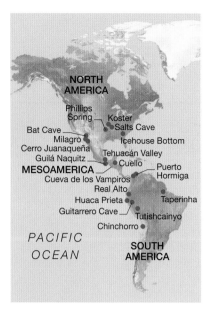

KEY SETTLEMENTS
Agriculture developed early in Mesoamerica and northern South America, later in eastern and southwest North America. In many other regions, a great variety of well-adapted forager lifestyles endured until they were fatally disrupted by the arrival of Europeans.

EVOLUTION OF CORN
Teosinte's spikes and tiny grains were gradually transformed into large corn ears bearing tightly packed plump kernels.

TEOSINTE LEAVES AND SPIKE

MODERN CORN COB

Mesoamerica

Foragers in Mesoamerica were sowing, planting, and tending selected wild plants, such as zucchinis and bottle gourds, by 8000 BCE, initiating domestication. The variety of these plants increased through time, as communities in different regions exchanged and spread local plants. By 4500 BCE these included corn. Nevertheless, for many millennia agriculture made only a small contribution to the diet, and communities still moved around to exploit seasonally available wild resources. By 1600 BCE, however, the size of corn ears had increased sufficiently to enable corn to become the staple food supporting permanent settlements. Protein was obtained by growing beans, hunting animals such as deer and wildfowl, and catching fish. Recently, botanical evidence has revealed that foragers in the tropical rain forests of the Isthmus and northern South America also developed horticulture at an early date, by 6000 BCE, growing a wide range of roots and tubers, such as sweet potato, zucchinis, beans, and treecrops, such as guava.

The Amazon

Early Amazonian settlement is known mainly from coastal shell middens associated with large villages. The occupants hunted land and marine mammals, gathered wild plants, fished, and, probably by 6000 BCE, cultivated some plants. Early agriculture in the interior is implied by indirect evidence, including plants grown in the Andean region that must have been domesticated in the Amazon, and microliths used in graters for processsing manioc, which became a staple Amazonian crop.

MANIOC TUBERS
Manioc yields highly nutritious starch but must first be grated, soaked, squeezed then roasted or boiled to remove deadly cyanide.

The Andean region

The Andes contain a range of contrasting habitats, from coast to high grasslands, spread over a relatively short horizontal distance. This led to the development of vertical economies, in which people exploited the resources of the different zones through vertical movement and trade. Early agriculture began in various parts of the Andean region after 8000 BCE, with crops including chilli peppers, beans, zucchinis, potatoes, and the cereal quinoa. Guinea pigs (for meat), llamas, and alpacas were domesticated in the high plateau by 5000 BCE. Seasonal pastoral camps in this region later became permanent settlements with crops. Exchange networks spread domestic plants between Andean zones and introduced many plants from the Amazon and Isthmus rain forests, as well as corn from Mesoamerica. Coastal dwellers depended mainly on the rich marine resources, but also cultivated plants in the rainless valleys of the coastal plain, including bottle gourds for fishing floats and cotton for making nets, fishing line, and textiles. Those in the south, around Chinchorro, mummified some of their dead, particularly children, perhaps so they could be displayed in rituals.

North America

Domestic corn, beans, and squash were introduced into the American Southwest from Mesoamerica after 2000 BCE. Irrigation channels and hillside terracing are known here by 1300 BCE. Later, some local plants began to be cultivated and other Mesoamerican crops were adopted. Turkeys were domesticated by around 800 BCE. In eastern North America, foragers exploiting the rivers, lakes, and woods also began to cultivate local plants after 2500 BCE. They traded extensively, circulating materials such as seashells and Great Lakes copper.

PREPUEBLO PIT HOUSE
Early farmers in the Southwest lived in villages of pit houses, roofed by wooden poles and mud, supported on central posts, with an internal hearth. Storage pits were located inside or out.

CHINCHORRO MUMMY
From around 5000 BCE, the Chinchorro hunter-gatherers mummified their dead. They removed the flesh; reinforced the skeleton; stuffed the body with plant and other material; replaced the skin, coated it in clay, and painted it.

VALDIVIA FIGURINE
The Valdivia fisher-farmers of coastal Ecuador were the first people in South America to make pottery, around 4500 BCE. They also created fine minimalist human figures in stone.

clay mask

stick support

plant packing

DOMESTICATED LLAMAS
Native camellids were of great importance to Andean communities. Llamas were invaluable as pack animals; alpacas were kept for their wool. Both provided meat and hides, as did their hunted wild relatives, the vicuña and guanaco.

Getting more from animals

Animals were initially kept for meat, hides, bones, and manure. Feeding animals on crop surpluses made them a food reserve, and large herds represented wealth and prestige. But domestic animals became far more important when people began using them also in other ways: for milk, wool, eggs, traction, and transport.

Animal products

Eating an animal ends its usefulness. Productivity increased greatly when people began harvesting renewable resources from living animals: milk from cattle, sheep, goats, camels, and horses; wool from sheep and alpacas; and eggs. Milking began in Europe during the 5th millennium BCE and probably much earlier in West Asia, and new forms of pottery associated with milk production appeared widely. Wild and early domestic sheep bore a winter undercoat of short wool, shed in the spring, that was useless for making textiles. Sheep bearing suitable wool were bred in West Asia in the 4th millennium BCE and spread quickly into Europe and Central Asia.

SOAY SHEEP
Primitive breeds, such as soay sheep, that shed their wool were plucked or combed; later breeds were sheared.

WOOL AND SPINDLE
Spindle whorls (the disk used to weight the spindle) of stone, shell, or pottery are common finds at archaeological sites. Spinning became a perpetual task for women once sheep were kept for wool.

EGYPTIAN MEN MILKING A COW
Egyptian, Saharan, and Mesopotamian art reveal ancient milking practices. Usually the cow's calf was kept beside her, so that she would let down her milk. Unlike these Egyptians, Sumerian farmers sat behind the cow to milk her.

Animal labor

Using animals to carry and pull vastly increased what people could achieve. Donkeys, cattle, llamas, camels and horses, and even dogs and sheep, were used as pack animals. The invention of the wheel further increased productivity as animals could pull wheeled vehicles with far heavier loads than they could carry. The plough, invented in the 5th millennium BCE, was drawn in the West by oxen and in the East by water buffalo. Donkeys, usually pack and draught animals, were sometimes ridden. Horses were prestige animals, mainly drawing war chariots; they were rarely ridden before the 1st millennium BCE, when saddles and sophisticated bridles were developed. Bactrian camels were used for meat, fur, traction, and transport in Asia. Arabian camels, perhaps kept initally for their rich milk, were important pack and riding animals by 1000 BCE, making it possible for traders and pastoralists to exploit the Sahara and Arabian deserts. In the Andes, trade and the interzonal transport of goods depended hugely upon the llama.

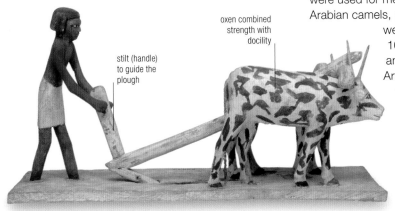

stilt (handle) to guide the plough

oxen combined strength with docility

OLD KINGDOM PLOUGHMAN
Ploughing enabled grassland and heavy ground to be broken up and greatly increased the area that a single farmer could prepare for cultivation. Possible but difficult for a man to draw, the plough was generally drawn by animals.

WHEELS AND WARFARE

The invention of wheeled transport in the 4th millennium BCE had many practical benefits, but warfare was a major stimulus to further development. To be visible to their troops, war leaders rode in chariots with four solid wheels, drawn by donkeys. In the early 2nd millennium BCE, chariots with two lighter, spoked wheels were developed, drawn by horses. By 1300 BCE, they had become the luxury vehicle from Egypt to China. Gradually chariots changed from elite transport to and from the battlefield into efficient mobile fighting platforms.

Colonizing the steppe

The Central Asian steppes contain extensive grasslands that offer good pasture but little else. Three developments made it possible for people to move into and exploit this region: milking, providing a renewable food supply; carts, to carry the people and all their possessions, including tents; and draught animals to pull them, initially oxen but later also horses. Horses were also ridden, allowing the nomads to herd their animals more easily and to travel rapidly when required; they also provided additional milk and meat. By the 1st millennium BCE, fully nomadic pastoral communities were spread right across the region. Steppe nomads often engaged in a little agriculture, planting crops at places to which they returned later to gather the harvest; they also obtained grain and other agricultural produce by trading with or raiding settled communities on the steppe margins.

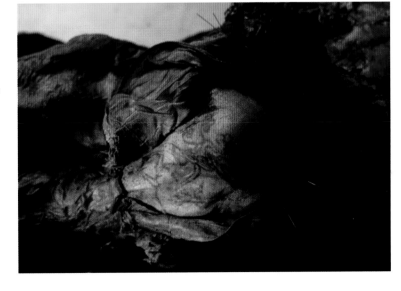

TATTOOED ICE MAIDEN
Many nomad burials in the cold Altai region froze in the ground, and bodies and artefacts of organic materials are often found perfectly preserved, like this heavily tattooed young woman, buried 2,400 years ago.

HERDING SHEEP
Steppe nomads moved frequently to keep their animals—sheep (see below), goats, cattle, or horses—supplied with fresh pasture. They did not wander at random but followed regular annual routes.

Craft development

Artifacts have played a critical role in human development, as tools but also as vehicles for artistic creativity. They bear witness to remarkable skills, patient labor, and technological discoveries, often built on previous achievements. For example, control of fire, initially for heat, cooking, and defense, led to ceramics, metallurgy, and glass.

Clay and sand

As well as having practical utility, pottery provided a versatile medium for artistic and cultural expression in its shapes and decoration. Surfaces could be painted, incised, impressed, cut away or roughened, or pieces of clay could be added and shaped. Pots were often built up of clay coils or slabs. Alhough pottery was often fired in simple bonfire kilns, more sophisticated kilns, with a firing chamber above or to one side of a fuel chamber, gave better control of firing conditions, and could achieve much higher temperatures; this technology benefited not only ceramics but also other industries such as metallurgy.

A related ceramic material, faience, was made from the 5th millennium BCE. A paste of silica sand, an alkali (such as potassium), and lime, colored with a metallic oxide (usually copper), was molded into objects that were then fired. The surface fused into a glasslike glaze, while the grains inside sintered (adhered). Glass was invented later, by 2200 BCE.

POTTER'S WHEEL
The potter's wheel, invented in West Asia around 3400 BCE, introduced mass production. The constantly turning wheel allowed uniform pots to be thrown at speed, in contrast to hand-building methods.

KARANOVO STAG
Imaginative vessel shapes were carried to extraordinary lengths by some cultures, creating drinking vessels, containers, and other pots in human or animal forms, like this Neolithic stag from the Balkans.

FAIENCE HIPPOPOTAMUS
Generally, only small objects were made of faience, such as tiny jars, seals, beads, pendants, and small figurines, such as this lovely Egyptian hippo. Gray areas show where the glazed surface has worn away.

Skilled craftsmanship

Many materials were worked in antiquity, including wood, leather, bone, horn, antler, ivory, and stone, often achieving astonishing results despite employing simple tools. Hard stone was shaped with a hammerstone, then ground and polished, using sandstone or wet sand as an abrasive, to its finished form as a tool, jewelry, or sculpture. Abrasives were also used when drilling beads, using stone or metal drill bits. Shell was popular for jewelry, such as beads and bangles, decoration on clothing, and inlays. Iron ores were also decorative materials: hematite (red ocher) was widely used as a pigment, while in Mesoamerica, magnetite and ilmenite were carved, ground, and highly polished as mirrors worn by the elite. An object's value could depend not only on the rarity and beauty of its material and final appearance but also on the skill, time, and effort invested by its maker.

BEADMAKING
Beads were made from many materials, such as bone and shell (as in this necklace from Palestine), and gemstones, of which chalcedonies such as agate and carnelian were particularly popular.

OLMEC JADE HEAD
Jade, an exceptionally hard stone, was highly valued by many cultures for sculptures, polished axes, jewelry, and other prestige objects.

Textiles

Early textiles were woven from plant fibers, including flax, hemp, and cotton. Wool was woven or felted, a technique known since the Bronze Age; knitting came later. Textiles were used for clothing but also for blankets, tents, and sails. Cloth was dyed with plant or mineral pigments. In South America, cotton cloth was sometimes embroidered using alpaca wool; by Chavín times (see p.242) decorative techniques included tapestry-work, tie-dyeing, batik, and painting. Many woven designs could be created by varying the interplay of warp and weft threads, in single or multiple colors. Textiles were hugely important in many societies, partly because of their variability.

THE LOOM
Many areas used a simple backstrap loom; to maintain tension on the warp (vertical) threads the free end was fastened to an upright (such as a tree). Ground (horizontal) looms were common in Asia. Upright looms, used in Europe, had weights attached to the warp threads, as depicted in this rock engraving at Val Camonica, Italy.

SILK

Forming the silkworm's cocoon, silk is a continuous fiber, up to 3,280 ft (1,000 m) long, which is broken when the moth emerges. But plunging the cocoon into boiling water (killing the larva) dissolves the gum that holds the fibers together, allowing the silk thread to be carefully reeled off. Silk, used in China by the 5th millennium BCE, became a highly sought-after commodity that stimulated the development of West–East trade.

Clothing and identity

Personal adornment has been important to humans for at least 100,000 years, but jewellery and decorated clothing became much more important in the last 10,000 years. These also served as markers of personal identity, indicating membership of a particular community, life stage (such as child, adult, married, elder), gender, and status. As societies became more complex and unequal, the variety of social roles increased, as did the range of markers needed to show these distinctions. Status and role were visible through such things as clothing, hairstyles, personal ornaments, and other objects, such as badges of office or work tools. These signs made people's social identity clear to those whom they encountered, facilitating decisions about behavior and interactions. Even in death, grave goods reflected a person's many roles in life.

DRESS CODE
This Old Kingdom scene contains many clues to social identity in Egypt. The workers wear simple kilts; more elaborate clothing, a necklace, a staff, and a scepter distinguish the steward; a (red) dress identifies a woman.

VARNA COPPER AGE BURIAL
In Bulgaria's Varna cemetery, dated 4500–4000 BCE, there are big differences among the 300 burials in the quantity and richness of the grave goods. This man was buried with 990 gold objects, including discs sewn onto his clothing.

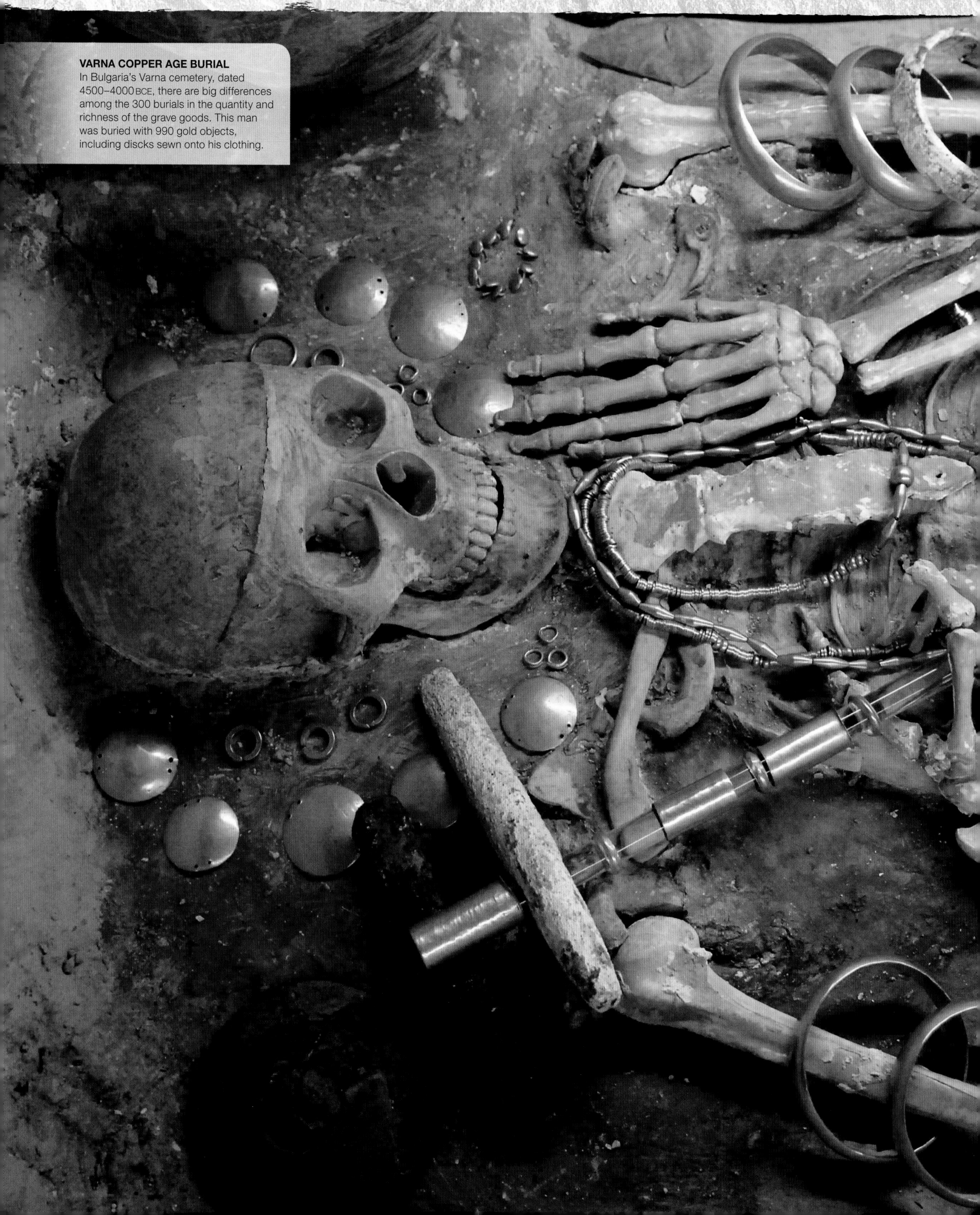

Metalworking

The early development of metallurgy was led by the social importance of metals for making prestige objects with which people could show their status. Only later, with the development of alloys, did metal also become significant as a material for tools and weapons.

From hammer to kiln

Pure nuggets of gold and copper were cold-hammered into shape in West Asia from before 9000 BCE. In eastern North America, from around 3000 BCE, the Old Copper Culture cold-hammered Great Lakes copper, making and trading small objects. Smelting copper and lead ores began in West Asia after 7000 BCE, and by the 6th millennium BCE kilns reaching temperatures high enough to melt metal made casting possible. By the 5th millennium BCE, simple molds producing flat shapes had evolved into elaborate molds for casting 3-D objects. For a long time, metal was used mainly as a display of wealth and importance.

EARLY CHINESE AXE
Axes in early China mainly served ritual purposes, such as human sacrifice.

Metal put to work

By around 4200 BCE, copper ores containing arsenic were being selected to produce an alloy harder than copper and therefore more suitable for tools. During the late 4th millennium BCE, tin began to be added to copper, producing a harder alloy: bronze. This was still used for prestige objects, but also now for tools. Metallurgy spread to East Asia around 3000 BCE, and multiple-piece molds were developed in China around 1700 BCE to create elaborately decorated vessels. In South America, gold and copper working began after 2000 BCE and sophisticated technology developed rapidly.

COPPER INGOT
Copper and tin were traded in the form of distinctive "oxhide" ingots in the Mediterranean during the 2nd millennium BCE.

MINING

Gold and tin could be obtained by panning but metals were generally mined. Depending on the nature and depth of the ore deposits, people dug open pits or cut shafts and radiating galleries, shored with timber supports. The rock was cracked by fire-setting, removed using antler picks and levers, stone hammers, and wooden wedges, and crushed with stone tools to extract the ore. Remarkable copper mines dating from the 5th millennium BCE have been found at Aibunar and Rudna Glava in the Balkans. Ores were often smelted close to the mine, although this required a good supply of timber.

Metallurgy in Europe

Copper smelting and casting and a remarkable gold-working tradition developed in the Balkans around 5500 BCE. By 2500 BCE, metallurgy had spread through Europe, associated in the east and north with the Corded Ware Culture and in the west with the Beaker Culture. Gold and copper jewelry and daggers were among many items used as status symbols. Bronze-working became widespread after 1800 BCE, when all parts of Europe were linked by trade routes circulating metals, particularly tin. Bronze jewelry, weaponry, and armor displayed their owners' power in the emerging hierarchical societies. The Minoan and Mycenaean civilizations of the Aegean are the best known, but rich barrow burials reflect the presence of chiefdoms elsewhere in Europe.

MYCENAEAN DAGGER
Inlaid in gold and silver with a marine design, this bronze dagger is from an Early Mycenaean elite burial in a Tholos tomb.

Trade

Settled communities have always needed trade mechanisms to obtain materials not available locally. As societies became more complex and demanded a greater range of commodities, trade became more controlled and efficient. Trade also brought about the spread of ideas, knowledge, and technological innovations, significantly affecting global development.

Why trade?

Hunter-gatherers and pastoralists could obtain many raw materials as they moved from place to place, but settled communities had access only to local resources, and had to find other ways to obtain more distant ones. Early farmers needed flint and hard stone for tools, and other materials, such as seashells and gemstones, for social purposes; later, metals were also required. As more intensive agricultural systems developed—supporting growing populations—demand increased for materials to serve an ever-increasing range of purposes. In addition, environments suited to intensive farming, such as major river floodplains, were often far from the sources of desired materials, many of which came from mountains.

HAFTED STONE AX
In Australian gift-exchange networks, stone axes from inland quarries moved by small steps over a wide area in exchange for such things as possum-skin cloaks.

COPPER MINE
The 3rd-millennium BCE development of copper mines in the Timna Valley (above) and other sites in southern Israel was probably connected to the growing demand for copper from Old Kingdom Egypt.

GOLD NECKLACE FROM UR
Many of the outstanding gold pieces from the Ur Royal Cemetery (see p.234) included lapis lazuli, as here. This necklace also has carnelian from India.

CHLORITE VESSEL
Chlorite (soapstone) from Iran's Kerman region was made into highly distinctive vessels that were widely traded in the 3rd millennium BCE.

OBSIDIAN

Obsidian (volcanic glass) was widely prized for its beauty and extreme sharpness when freshly knapped. In Neolithic West Asia, gift-exchange carried obsidian more than 500 miles (800km) from several sites in Turkey. Communities near the sites made most of their tools from obsidian.

Trade networks

Although they could travel great distances, desired or essential raw materials moved initially in a relatively haphazard fashion, through gift-exchange between neighboring communities. Such gifts, presented in the context of formalized exchange partnerships or social ceremonies such as marriage, served social as well as economic purposes: givers might seek prestige through the generosity of their gifts, rather than material returns.

As larger and more complex societies emerged, with growing demands for materials, mechanisms developed that could ensure a more reliable and controlled supply. Those who wielded power made contact with communities near source areas or on routes along which goods moved, entering into trade agreements or setting up trading enclaves within these communities. Having obtained the materials, they redistributed them among their followers or used them to enhance their own status or that of their community or deity. With the emergence of states, these trade arrangements grew and developed: materials were traded in far greater quantities, with greater regularity; and trade also employed techniques such as political coersion or conquest, expeditions to sources, and the establishment of permanent trading colonies.

Map labels:
Arslantepe
Tell Brak · Nineveh
Habuba Kebira
Mari
Eshn___
Khafajah
Sippar
SYRIAN Kish
DESERT Shuruppak
Umma
Uruk
Ur
Red Sea

TRADE IN WEST ASIA

— Trade routes
● Towns and settlements

COMMODITIES

▲ Copper	▲ Silver/Lead
▲ Tin	▲ Gold
▲ Lapis lazuli	▲ Shell
▲ Chlorite	▲ Turquoise

WEST ASIAN TRADING TOWNS
The growing trade demands of towns in West Asia in the late 4th millennium BCE stimulated the development of a network of trading towns across the Iranian plateau, located near source areas or at key places along the routes.

FARMERS AND TRADERS
Throughout the 4,000 years of the settlement's existence, people at Mehrgarh (Pakistan) were able to obtain materials such as copper, soapstone, turquoise, lapis lazuli, and shell from distant sources.

TURQUOISE
By the 5th millennium BCE, turquoise from southern Turkmenia reached not only Mehrgarh but also West Asia.

LAPIS LAZULI
Probably the most highly prized traded commodity, the only known source of this beautiful blue stone was Badakshan.

MODES OF TRANSPORT

Walking while carrying loads was the original form of land transport and the only one in regions without pack animals, such as Mesoamerica. Wheeled vehicles could usefully move large loads over short distances but generally not over long distances before roads were built. Pack animals were used to transport relatively light goods such as textiles or high-value, low-bulk goods such as tin or silver, which people could also carry. Water transport was often preferable, being generally cheaper and quicker, and it was essential for bulk transport of heavier materials. Boats, rafts, and inflated skins were used on rivers; wood could be towed or floated downstream. Sea transport was also used where possible, despite natural hazards such as currents, storms, rocks, and shoals. The discovery of shipwrecks, such as the Uluburun vessel, gives a fascinating insight into ancient trade networks.

REIN RING WITH DONKEY
This silver rein ring was part of a sledge from the Ur Royal Cemetery, drawn by a pair of oxen. Donkeys were the main pack animals in West Asia.

INDUS VALLEY CART
The Indus people used various types of cart, drawn by bullocks, to transport farm goods, or as with this example, as passenger vehicles in towns.

ULUBURUN REPLICA
The original ship, on a voyage around the eastern Mediterranean, was wrecked off Uluburun in Turkey around 1300 BCE. Its cargo included copper, tin, and glass ingots.

LEAD/SILVER
By 3800 BCE in West Asia, silver could be extracted from smelted lead ores. Silver later became the main medium of exchange.

TIN
Since tin was rare, its use in bronze from the late 4th millennium BCE was a major stimulus to the development of trade networks.

INDIAN CHANK SHELL
Seashells, obtained by coastal fishing communities, were distributed far inland by gift-exchange from Neolithic times.

Caspian Sea

BADAKSHAN

IRANIAN PLATEAU

Godin Tepe

Tepe Sialk

Tepe Hissar

a

gash

Anshan

Tal-i Qal'eh

Shahdad

Tal-i Iblis

Tepe Yahya

Konar Sandal

Bampur

Shahr-i Sokhta

Mehrgarh

INDUS VALLEY

THAR DESERT

Persian Gulf

Tarut

Umm-an-Nar

Arabian Sea

ARABIAN PENINSULA

Religion

Religion was integral to ancient people's understanding of their world and was inextricably entwined with much of their daily lives. Although without written records their belief systems are unknowable, monuments, artifacts, and art provide clues to their religious practices.

RAINBOW SERPENT
An important being in Australian creation myths, the Rainbow Serpent is said to have forced her way out of the ground and across the landscape, creating rivers and gorges and giving rise to many tribes.

The role of religion

Prehistoric religion reflected people's need to understand the world, ensure natural and social order and prosperity, and explain disasters beyond human control. Through rituals and offerings ancient societies sought to bribe or appease the divine forces controlling the world or its individual components. Many had shamans or priests who mediated between people and the gods and often enjoyed power within their community. Creation myths and stories of the gods explained how the world began and functioned. The form of religion in different societies has varied greatly, with many different deities, in human, animal, or inanimate form, including ancestral spirits. Ways of worshipping included music, dance, processions, costumes, formulaic recitation, offerings or sacrifices, and consuming drugs, ritual foods, or drinks.

KING WITH GOD
The divine Egyptian king was both a secular ruler, interceding with the gods on his people's behalf, and the gods' representative on Earth. Here Seti I stands before the hawk-headed god Horus.

Death and disposal

Since Neanderthal times, people have practiced rites that showed concern for their dead, perhaps linked to belief in an afterlife. Burial, in graves or tombs or under house floors, was common, but many societies practiced other rites, including cremation, exposure, or disposal in watery places. Some thought it important to preserve the body, so they undertook mummification (for example, in Egypt and South America). Monumental tombs, such as tumuli, pyramids, and megaliths, could link the living and the dead to ancestral lands or sacred places; these might house single individuals, families, social groups, or whole communities. Grave goods accompanying the dead could reflect their roles in life, such as gender, age, status, and occupation, but also related to religious beliefs, such as providing food for use in the next world.

SHANG CHARIOT BURIAL
Horse-drawn chariots, introduced into China from the steppe around 1300 BCE, were used by the Shang elite in warfare and hunting and were buried with them as a mark of their high status.

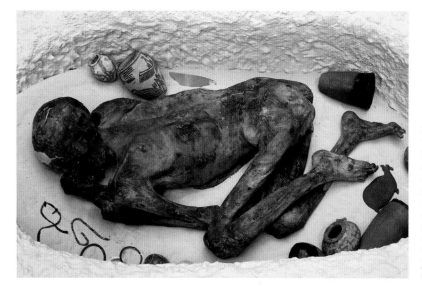

SAND MUMMY
"Ginger," an Egyptian buried in sand 5,000 years ago, was preserved as a natural mummy by the hot dry conditions. Placed around him are objects for his use in the next world: pots, jewelry, a stone jar, and a stone palette.

Shrines and temples

Sacred places, where the material and spiritual worlds met, took many forms. While landscape features believed to house or embody deities, such as woods, rivers, caves, or mountains, were often places of worship, many societies built a variety of shrines, temples, and religious monuments, from household niches to vast sacred complexes, often housing an image of a god. Here, people could communicate with their deities, in person or through a priest. Many sacred places were intended to inspire awe through their scale and magnificence and they were often decorated with religious sculptures or paintings. In some societies, sacred places were designed so that the entire community could participate in rituals; in others, access was restricted to officiating priests.

traces of black or green paint

wide eyes focused on the gods

EYE IDOLS
Hundreds of tiny "eye idol" figurines were found in the 4th millennium BCE temple at the Mesopotamian town of Tell Brak, representing individuals (left), mothers and children (far left), or families. They were probably votive offerings, asking for the god's favor.

CHAVIN LANZON
Worshippers braved eerie noises in the passages under the Chavín de Huantar temple in Peru to reach this terrifying image of the god.

ZIGGURAT OF UR
The temple raised toward heaven on this high tower provided a portal through which Ur's god Nanna could descend to Earth.

Astronomy

Many prehistoric groups developed an intimate and sometimes astonishingly accurate knowledge of the movements of the Sun, Moon, planets, and stars, from which they could establish the length of the year and the timing of important dates within it, such as the equinoxes and solstices. This calendric information was vital as the basis on which important seasonal decisions were taken, such as when farmers should plant crops. Since heavenly bodies were often regarded as visible manifestations of gods, observing their behavior was one of many ways used by priests and shamans to predict or interpret the will of the gods (others included various methods of divination and utterances in a drug- or dance-induced trance). The heavens were observed in conjunction with landscape features that provided fixed markers, but people also created markers, such as the spectacular stone circles and tombs of prehistoric Europe.

STONEHENGE
Begun around 3100 BCE, Stonehenge underwent many changes before the final arrangement of its stones around 1500 BCE. It probably served many purposes through time, including marking the midsummer sunrise.

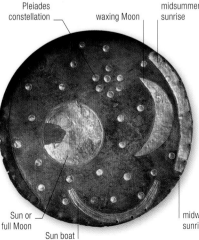

Pleiades constellation

waxing Moon

midsummer sunrise

NEBRA SKY DISK
This unique object, buried on a German hill around 1600 BCE, was probably an instrument for astronomical observation used in conjunction with the surrounding landscape.

Sun or full Moon

Sun boat

midwinter sunrise

Newgrange

The magnificent passage grave at Newgrange was one of a few Late Neolithic monuments in the western British Isles that combined the final flowering of the long tradition of megalithic tombs with the new idea of building stone monuments with an astronomical significance: the midwinter sunrise strikes through its façade to its heart.

Newgrange lies in the Boyne Valley, County Meath, Ireland, within a major Neolithic ritual landscape that also contains two passage grave mounds of similar size at Knowth and Dowth, many smaller passage graves, tumuli, standing stones, and enclosures. Situated in a dominant position on top of a hill, the monument is a huge circular mound built of stones and turves with a kerb of 97 massive slabs around the outside. On the southeast side, the mound is recessed into a stone façade leading into the cross-shaped chambered tomb (passage grave) covered by the mound. It was constructed around 3250 BCE.

Megalithic (meaning "large stone") tombs of many forms were built mainly in regions of Atlantic Europe during the Neolithic period, beginning in the 5th millennium BCE.

A few were freestanding; most were covered by a mound of earth or pile of stones (cairn) many times their size. Generally they housed the bones of many individuals, often collected after exposure, although rites varied considerably. In Britain, most were long barrows covering chambered tombs, but from the late 4th millennium BCE, passage graves under round mounds appeared in Wales, northern Scotland, and Ireland. These marked the beginning of a new tradition of arranging stone settings, such as circles, avenues, and alignments, to record astronomical events, such as the solstices. They also belong to the final stage of megalithic burial; during the later 3rd millennium BCE, this was replaced by burial in single graves, often under round barrows.

1. Decorated kerbstone. Geometric designs, including circles and spirals, were carved on passage graves in Ireland and Brittany, and elsewhere, suggesting shared iconography and belief systems.
2. Aerial view. The mound's kidney-shaped form is clearly visible from the air, as are the remains of a circle of large standing stones (orthostats) that once surrounded the mound.
3. Threshold stone and entrance. This massive decorated stone blocks the entrance, above which is the "roofbox," a hole constructed at the exact position to admit the sun at the midwinter sunrise.

4. Newgrange mound. While the internal structure is much as found, the wall-like façade is a fanciful reconstruction, built from original stones that had tumbled out of position over the ages, but reinforced with concrete.

5. Megalithic passage. The long passage of massive stones (orthostats) infilled with drystone walling led to a central chamber with three side chambers, containing stone basins that once held cremated remains.

6. Chamber ceiling. The roof of the central chamber was corbel-vaulted: each stone was laid to overlap the one below and the final space was closed by a capstone.

6 **5**

The first states

Intensive agriculture supported dense populations and produced surpluses used for trade and to support craftsmen and leaders. As societies became more complex, social control, religious authority, craft production, and other nonagricultural activities became concentrated in centers of population where rulers resided, creating cities.

Getting more from the land

Intensifying agricultural productivity allowed more people to be supported from the produce of a single area. Chief among the methods devised was irrigation, which could increase yields in farmed areas and make agriculture possible (and often highly productive) in areas with inadequate rainfall, for example. Water conservation and drainage were also important. Other methods included hillside terracing, using fruit trees to shade vegetables from excessive heat, and experimenting with new crops. Agricultural intensification often encouraged population growth and produced enough to support specialists, such as priests and artisans. Surpluses were also traded, not only for essentials but also for precious materials to decorate palaces and temples. Growing populations brought increasing scope for problems and so religious and secular leaders emerged to manage both interactions within the community and external relations. The process of growth culminated in the emergence of complex societies in the 3rd and 2nd millennia BCE in the Old and New Worlds.

ANCIENT HILLSIDE TERRACES
Systems to promote agricultural productivity began in Yemen well before 3000 BCE, including simple dams in the dry lowlands and terraced fields to maximize the fertile hillside land where rain fell.

The city

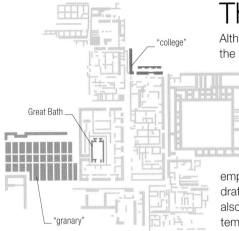

"college"

Great Bath

"granary"

PLAN OF THE CITADEL
State buildings on the citadel at Mohenjo-daro included the Great Bath, a carefully secluded pool, probably used for ritual purification.

Although early states depended upon the produce of the countryside, cities were their heart. The city was the territorial capital, to which all routes led and from which a network of towns and villages spread out. It was the state's political and administrative center, where those in authority resided, and the economic center, where taxes and traded goods were collected, stored, and disbursed as payments to state employees, such as artisans, or as rations to people drafted to work on public works. The city was also the religious center, housing the gods' chief temple; the artistic center, with all the finest buildings and public works of art; the intellectual center; and the population center, where many of the state's inhabitants lived.

Of course, not all cities fulfilled all these functions, but most fulfilled many. The state was often seen as belonging to the gods, for whose benefit the people labored, and who in turn cared for the people, but in practice the main beneficiaries were generally the rulers. Nevertheless, their religious beliefs meant they usually felt obliged to fulfill their duty of care for those they ruled.

COVERED DRAIN
A sophisticated water supply and drainage system was a distinctive feature of the Indus civilization in Pakistan and northwest India. Many houses at Mohenjo-daro had a private well, and most had a bathroom and a toilet.

PRIEST-KING
The political organization of the Indus state is still a mystery, since clues are few. This tiny sculpture, just 7 in (17.5 cm) high, is often taken to represent a ruler, but might equally portray a god. Indus stone sculptures were rare.

States

The first states took various forms. City-states were united by a shared culture, but each acted independently, controlling the surrounding towns, villages, and farmland, and allying with or fighting its neighbors. The first West Asian states were of this form. Territorial states were much larger, centered on a metropolis, with a hierarchy of lesser cities, towns, and rural communities. Some, like the Akkadian empire (see p.234), directly controlled only the heartland and key external locations, such as mines. In others, state control was universal. There were also complex societies united by a common culture and, more importantly, a common religion, who acknowledged the preeminence of a cult center that had many features of a city, but who were not politically united; this may have been true of the Olmec and the Chavín in the Americas (see p.242).

DUCK WEIGHT
State bureaucracy and efficient management often involved standardization. Weights conforming to an official standard are well known from Mesopotamia, where this duck shape was popular.

Writing systems

Most early states developed writing systems, for different reasons. For example, the Sumerians (see p.234) first used writing for economic records, but the Shang in China (see p.240) wrote questions to the gods. Writing systems began with symbols to which meanings were attached. In West Asia and Egypt, logographs, representing words, were supplemented and gradually replaced by signs that signified sounds, usually syllables. During the 2nd millennium BCE, the alphabet was invented in this region. In East Asia and Mesoamerica, the mixture of logographs and syllabic signs remained much more complex.

CUNEIFORM TABLET
Many Mesopotamian clay tablets were fired, preserving a wonderful treasury of literature and documents of all kinds.

PICTORIAL SYMBOLS TO CUNEIFORM

Many early Sumerian signs were pictures symbolizing a word, like the three shown here. Around 3000 BCE, the writing tablet, and signs, were turned on their side. Drawing pictures with a reed stylus was laborious— by 2400 BCE the pictures were reduced to wedge-shaped (cuneiform) lines, later further simplified.

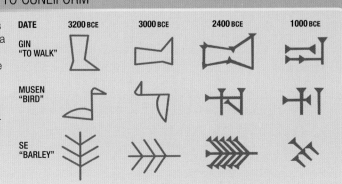

DATE	3200 BCE	3000 BCE	2400 BCE	1000 BCE
GIN "TO WALK"				
MUSEN "BIRD"				
SE "BARLEY"				

Mesopotamia and the Indus

In the 3rd millennium BCE states emerged in southern Mesopotamia (Sumer and Akkad) and the Indus region. Sea trade between them involved the cultures of the Gulf, such as the Magan, at the expense of the earlier land trade network. Texts, from bureaucratic records to epic poetry, reveal Sumerian life in detail. Far less is known for certain about the Indus civilization.

Sumer and Akkad

In the 4th millennium BCE, towns emerged in the rich lower Euphrates floodplain, supported by irrigation agriculture, and in the earlier 3rd millennium (Early Dynastic period) city-states developed. Each city housed many of the state's inhabitants and controlled the surrounding agricultural territory; animals were pastured in the uncultivated land between states. Although they were united by a common culture and acknowledged spiritual allegiance to the holy city of Nippur, the city-states were often in conflict with each other, and many built defensive walls. From 2334 BCE, Sargon of Akkad united southern Mesopotamia, creating an empire whose influence stretched to eastern Anatolia. This crumbled by 2192 BCE, after which some city-states enjoyed renewed prosperity. In 2112 BCE, the Third Dynasty of Ur (Ur III) reunited the region. Smaller than the Akkadian empire but exercising much tighter bureaucratic control, Ur III saw the greatest flowering of Sumerian culture. Ziggurats (see p.229) were built at Ur and other cities. The third king, Shulgi, issued the first Law Code. Texts describe a confident and well-ordered society, with a rich mythological tradition. This ended in 2004 BCE when the Elamites invaded and sacked Ur.

SUMERIAN RUINS AT URUK
By 2900 BCE, Uruk was a city, with perhaps 60,000 inhabitants. Most excavations have been in the Eanna precinct, dedicated to the city's goddess, Inanna, where many different religious buildings were uncovered, including this one, the purpose of which is unknown.

The Indus civilization

During the 4th millennium BCE, farmers and pastoralists from Baluchistan established settlements on the Indus plains, an area of great agricultural potential but prone periodically to devastating floods. Around 2600–2500 BCE many Indus towns were abandoned or demolished and replaced by planned settlements. Some, such as Mohenjo-daro (see pp.232–33), included massive brick platforms, raising all or part of the settlement above the maximum flood level. Generally, a part of the settlement—the "citadel"—contained public buildings, such as the Great Bath at Mohenjo-daro and the warehouse at Lothal. The lower town had well-appointed courtyard houses, often with an upper story, and craft workshops. At Mohenjo-daro, the largest city (with a population thought to be around 100,000), craft products included pottery, beads, silver vessels, copper tools, shell inlays, stoneware bangles, faience figurines, and exquisite soapstone seals.

EARLY FIGURINE
Human and animal figurines were among the most characteristic products of the Indus. This unusual early example comes from Harappa.

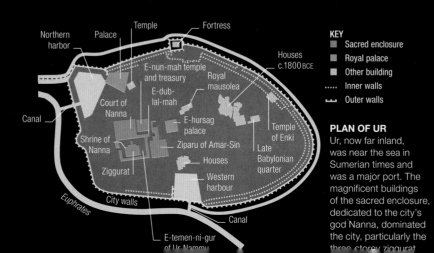

PLAN OF UR
Ur, now far inland, was near the sea in Sumerian times and was a major port. The magnificent buildings of the sacred enclosure, dedicated to the city's god Nanna, dominated the city, particularly the three-storey ziggurat

Northern harbor · Palace · Temple · Fortress · Houses c.1800 BCE · E-nun-mah temple and treasury · Royal mausolea · E-dub-lal-mah · Court of Nanna · E-hursag palace · Temple of Enki · Shrine of Nanna · Ziparu of Amar-Sin · Late Babylonian quarter · Ziggurat · Houses · Western harbour · Canal · City walls · Euphrates · Canal · E-temen-ni-gur of Ur-Nammu

KEY
- ■ Sacred enclosure
- ■ Royal palace
- ■ Other building
- ..... Inner walls
- ╍ Outer walls

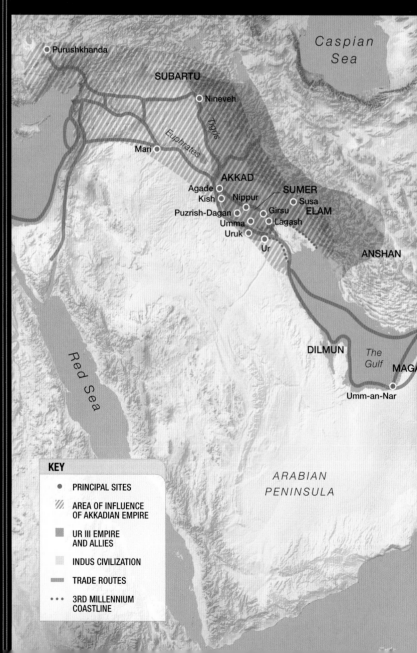

Purushkhanda · Caspian Sea · SUBARTU · Nineveh · Tigris · Euphrates · Mari · AKKAD · Agade · Kish · Nippur · SUMER · Susa · Puzrish-Dagan · Girsu · ELAM · Umma · Lagash · Uruk · Ur · ANSHAN · Red Sea · DILMUN · The Gulf · MAGA · Umm-an-Nar · ARABIAN PENINSULA

KEY
- ● PRINCIPAL SITES
- ▨ AREA OF INFLUENCE OF AKKADIAN EMPIRE
- ■ UR III EMPIRE AND ALLIES
- INDUS CIVILIZATION
- TRADE ROUTES
- ••• 3RD MILLENNIUM COASTLINE

Craft production was a major activity throughout the state. Specialist villages were set up to exploit local concentrations of resources, such as gemstones for beadmaking and shells for banglemaking in the coastal region of Gujarat. Lothal, a small town here, was heavily industrialized; it may have been manufacturing goods for exchange with local hunter-gatherers, who were well placed to obtain resources from the adjacent Thar Desert and Deccan Plateau, such as ivory, wild silk, wax, and honey. Gujarat was at that time partly islands, due to higher sea levels. The island city of Dholavira was probably the regional center, managing local and overseas trade, industry, redistribution, and taxation.

Nothing is known of the political organization of the Indus civilization, but the appearance of affluence in the towns and cities, the efficient distribution of craft products and other commodities (such as dried marine fish, found 530 miles/850 km inland at the city of Harappa), the standardized weights, and considerable cultural uniformity, show that this was a highly organized state. Despite this, around 1900–1800 BCE, it disintegrated. Towns and cities were abandoned or occupied by people in makeshift, inferior houses; writing ceased; and cultural uniformity was replaced by regional diversity, with flourishing rural farming communities developing in Gujarat and the eastern edge of the former Indus realm. Various factors probably contributed to this change: a reduction in the flow of the Saraswati, the important second river; the decline of sea trade after Ur fell; a changing agricultural regime due to the cultivation of new crops (rice and millets); and diseases such as malaria and cholera.

BEAD NECKLACE
The Indus people were highly skilled at manufacturing beads from a wide range of gemstones and other materials: this necklace includes tiny beads of gold.

serpentine bead

typical "etched" carnelian bead

Trading states

Mesopotamia had far-flung trade networks. Silver and timber probably came from Anatolia and the eastern Mediterranean. Mesopotamian texts record that ships docked in their ports from Meluhha (the Indus), Magan (the Oman Peninsula) and Dilmun (Bahrain). Dilmun was an entrepôt: besides excellent pearls and dates, it could offer sweet (fresh) water to visiting ships. Magan had rich deposits of copper and diorite, much prized in Mesopotamia for making statuary. But most of Sumer and Akkad's imports came from the Indus or were carried in Indus ships. These included timber from Gujarat and the Himalayas, for building boats, chariots, and furniture; copper from the Aravalli Hills, the Himalayas, and Afghanistan, source also of gold and tin; carnelian and other gemstones from Gujarat; ivory; and, surprisingly, lapis lazuli. This still came from Badakshan (see also p.227). It was not a material that the Indus people found particularly attractive, preferring harder stones that could retain a high polish. Nevertheless, they established a trading colony at Shortugai, 620 miles (1,000 km) to the north, from which they monopolized the supply of lapis lazuli. This indicates the importance they attached to trade with Mesopotamia. Little is known of what the Indus people imported in return, although a strong possiblity is woolen textiles, which the Sumerians manufactured on an industrial scale.

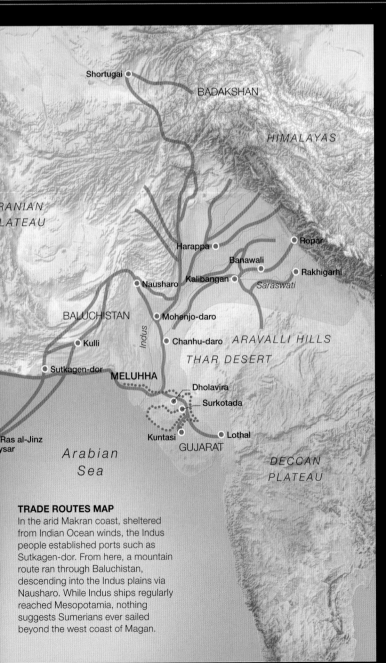

TRADE ROUTES MAP
In the arid Makran coast, sheltered from Indian Ocean winds, the Indus people established ports such as Sutkagen-dor. From here, a mountain route ran through Baluchistan, descending into the Indus plains via Nausharo. While Indus ships regularly reached Mesopotamia, nothing suggests Sumerians ever sailed beyond the west coast of Magan.

five Indus signs

INDUS SEAL
Indus soapstone seals, miniature masterpieces of craftsmanship only a few inches wide, bore an animal design and a short text in the undeciphered Indus script. The most common design was a unicorn; the zebu (pictured here) was rare.

Standard of Ur

Excavations south of Ur's Sacred Enclosure revealed a large Early Dynastic cemetery, where several graves contained objects inscribed with names matching those of early kings in the *Sumerian Kinglist*. The rich grave goods included the magnificent Standard of Ur.

Most people in the cemetery were interred in graves. Offerings buried in one included the gold helmet of King Meskalamdug, in the form of a wig, with every lock of hair lovingly modelled. However, 16 burials, known as the Royal Graves, are much grander: vaulted brick or stone tombs, containing people apparently willingly sacrificed to accompany the deceased. The largest group contained 68 court ladies and six soldiers. Royal grave goods included headdresses of gold leaves; a gaming board inlaid in lapis lazuli, red limestone, bone, shell, and red paste; and lyres decorated with a bull's or cow's head in gold and lapis lazuli. The "Standard of Ur" may, in fact, have been a lyre soundboard. It bears two scenes: war, followed by celebration of the victory.

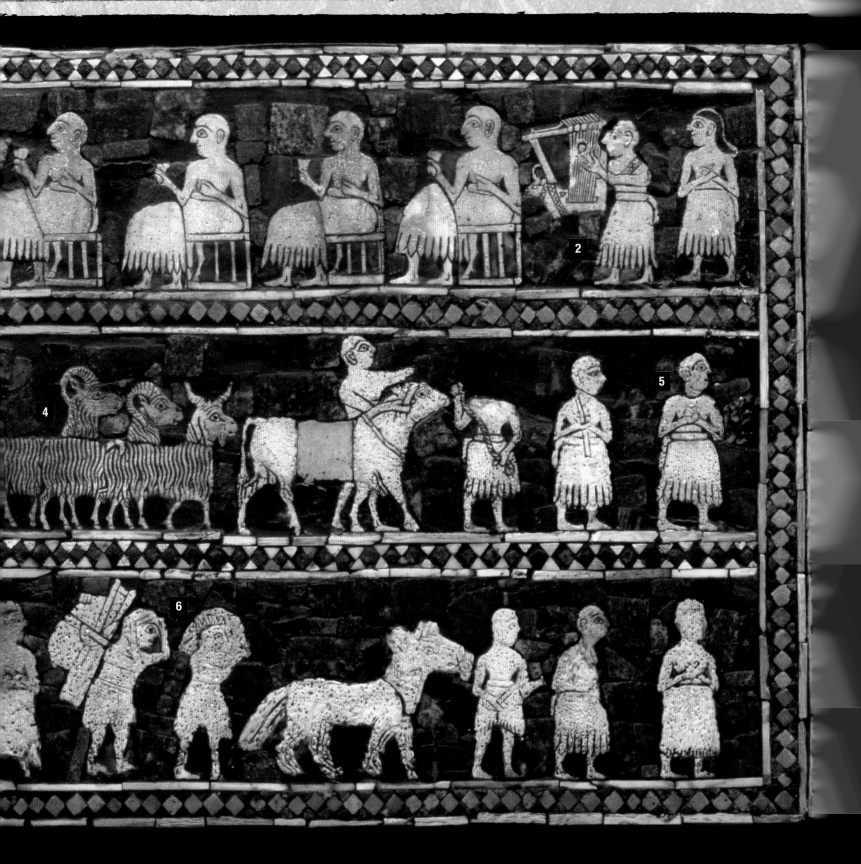

WAR SIDE

The story begins on the War side. At the bottom, Sumerian chariots carrying a driver and a spearman drive over the enemy dead. In the middle, the infantry march in a disciplined line; an enemy is captured, another falls, while others limp away. At the top, enemy prisoners are brought before the victorious Sumerian king and his court, while attendants unharness the royal chariot.

1. Seated ruler. The king, dressed in a distinctive woolen kilt, presides at the feast to celebrate the military victory.

2. Entertainer. A musician plays a lyre that is identical to animal-headed examples recovered from the Royal Graves.

3. Man carrying fish. Fish had been eaten, and offered to the gods, since farmers first settled on the Euphrates plains.

4. Woolly sheep. All the bounty of the land is being brought for the feast, including cattle, sheep, and goats. Sheep were kept mainly for their wool, to make textiles.

5. Procession leader. Each group is led by a man with his hands clasped on his chest, the position adopted in prayer.

6. The defeated. Sumerians guard war captives (distinguishable by their clothing style), who carry booty taken in the war.

Dynastic Egypt

A unified state emerged in Egypt around 3100 BCE. Unrestricted wealth and power allowed pharaohs to mobilize the enormous manpower and resources to build the pyramids. Under later dynasties, this centralized power was gradually dissipated and the state collapsed in 2181 BCE.

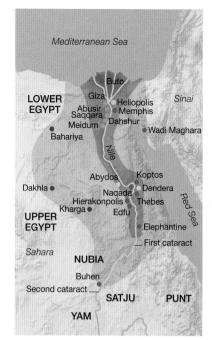

EXTENT OF OLD KINGDOM
The Egyptian state occupied the ribbon of land annually flooded by the Nile and was flanked by desert—source of building stone, gemstones, and some metals. Oases on the west also offered agricultural land and resources such as natron, a mineral used in mummification. Gold was sourced from the south, lumber from Lebanon in the east.

Unification of Egypt

The fertile Nile Valley supported prosperous farming settlements. Predynastic cemeteries in Upper Egypt bear witness to a developing social hierarchy and the importance already paid to the treatment of the dead. By 3500 BCE, there was growing cultural uniformity throughout the region, foreshadowing political unification as this cultural identity spread into Lower Egypt. Towns emerged in Upper Egypt, including Naqada, Hierakonpolis, Koptos, and Abydos. Competition and conflict between them encouraged the rise of leaders with both religious and secular power, who controlled the surrounding lands. Around 3100 BCE, a state emerged, centered on Hierakonpolis, uniting the lands of Upper and Lower Egypt. Writing, which developed around 3200 BCE, now became more widespread and more extensively used.

The Early Dynastic Period, dated 3100–2686 BCE, covers the time of the 1st and 2nd Dynasties (although some scholars include the 3rd Dynasty). A new capital was founded at Memphis, and the 1st-Dynasty rulers built mud-brick royal tombs at Abydos. Top officials were buried with copper objects and stone vessels in tombs at Saqqara, site of the 2nd Dynasty royal tombs. Cemeteries with wealthy burials found in many other areas show that power was still relatively spread throughout the kingdom.

NARMER PALETTE
Discovered at Hierakonpolis, this stone palette depicts King Narmer, the first Egyptian king (possibly identical to Menes, who united Egypt), smiting his enemies.

PYRAMIDS AT GIZA
Pyramid construction reached its peak with the 4th-Dynasty pyramids at Giza. The largest is the Great Pyramid of Khufu, which was 481 ft (146.5 m) high and contained 2.3 million stone blocks. The others here were built by Khufu's son Khafre and grandson Menkaure.

The Old Kingdom

Pyramids were invented and built during the Old Kingdom (2686–2181 BCE). The first was the Step Pyramid at Saqqara, designed by the architect Imhotep for Djoser, second king of the 3rd Dynasty. Built of six tiers of stone blocks, it stood 200ft (60m) high. Passages leading to the royal burial chambers lay beneath it. State organization was now highly centralized. At its head was the divine pharaoh, invested with enormous power, but also the huge responsibility of mediating with the gods to ensure the safety of the realm and the reliability of the natural world, particularly the annual flood upon which Egypt's prosperity depended. The pharaoh delegated power to high officials, originally from the royal family, who often exercised both secular and priestly authority. Estates granted to them for their maintenance were in principle held during their term of office but gradually became hereditary, creating a powerful and independent nobility.

Quarrying and transporting the stone for the huge pyramids and assembling and feeding the workforce to build them required enormous administrative and economic resources. Pyramids became smaller after the 4th Dynasty, suggesting the state's power was weakening. By the late 6th Dynasty, centralized control was breaking down and provincial governors were treating provinces as their own territory. There were probably also disastrously low Nile floods and famine.

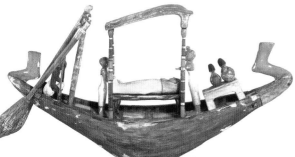

FUNERARY BARGE
Model boats were placed in tombs to symbolize the deceased's journey to Abydos, which was the center of the cult of Osiris, the god of death and resurrection.

SENEB AND FAMILY
Seneb, who lived around 2500 BCE, was the overseer of the palace dwarves, priest of the funerary cult of king Khufu and his successor, and chief of the royal wardrobe. His wife Senetites was a court lady and priestess.

FISHING SCENE
Fish were an important supplement to the peasant's regular diet of bread, beer, and vegetables, but fowling and fishing were also favorite sports of the elite, as in this scene from the tomb of Ti, a 5th-Dynasty nobleman.

Shang China

The Shang state in northern China is well known from historical sources and archaeological excavations of Shang cities and burials. Recent discoveries, however, show that historically unknown states of equal wealth and comparable development may have existed to the south.

Bronze Age China

Grave goods from cemeteries of the 3rd-millennium BCE Longshan culture provide evidence of growing social inequality in the Yellow River region. Several significant developments for the later Shang culture took place during the Longshan period. Between 2500 and 1900 BCE, divination using inscribed oracle bones began. Human sacrifices were now buried as foundation offerings beneath houses and walls. Jade and pottery making became specialized industries. Towns emerged, with elite housing and defensive walls constructed of *hangtu* (pounded earth); weaponry also indicates growing social tensions.

The emergence of a state in northern China is traditionally attributed to the Xia dynasty, dated around 1700–1500 BCE. Until recently semilegendary, information on the Xia is now coming to light, following the identification of Erlitou as their capital city. Occupation in this 740 acre (300 ha) settlement began in Longshan times. A step-change occurred with the appearance here of two palaces and craft workshops making stone, bone, and bronze artifacts—the first evidence of metallurgy in the region. Bronze vessels accompanied the elite, buried in wooden coffins.

XIA MONSTER
This bronze decorative plaque inlaid with turquoise, in the form of a monster with bulging eyes, was worn on the chest of an elite individual buried at Erlitou.

Shang dynasty

The Shang dynasty followed the Xia. Three successive Shang capitals, Zhengzhou (see right), Xi'ang, and Anyang, are known, with Anyang becoming the final capital around 1300 BCE. The central portion of the 10-square-mile (25-sq km) urban area was a complex of palaces, halls, and temples. Many had foundation sacrifices of animals and people, including chariots with horses and charioteers. Other parts of the city contained cemeteries, housing, and workshops. Craftsmanship and fine materials were applied almost exclusively to ritual and prestige objects, such as bronze vessels for feasting, ceremonial jade weapons, and bronze sacrificial axes; tools were made of stone, shell, bone, or wood. A great royal cemetery lay in the city's north, with 12 enormous pit tombs. Although robbed, 1,200 decapitated human sacrifices show the scale of their funerary provision. The tomb of Fu Hao, queen and general, in the city's southeast, was undisturbed. Her burial pit contained nested lacquered wooden coffins, 16 sacrificed adults and children, and more than 1,600 objects, including jade figurines.

ORACLE BONE
Questions to the gods were written on ox shoulder blades or turtle shells. Patterns of cracks induced by heat revealed their answer.

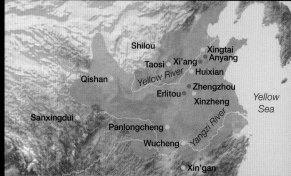

SHANG DYNASTY CHINA
Shang kings directly controlled the heartland around their successive capitals, while trusted nobles administered land granted to them elsewhere within the Shang realm. A ring of allied chiefdoms separated the Shang kingdom from enemy states.

KEY
- Successive capitals
- Other sites
- Extent of Shang influence

Oracle bones reveal that Shang kings consulted their ancestors and the supreme god Di for advice on many subjects. Despite the texts' brevity, they contain invaluable information on court life. The king frequently moved around the kingdom with his court, including senior officials, military officers, and priests. Trusted male or female former courtiers governed peripheral regions, defending the borders, raising taxes, and supplying manpower. Wars provided captives for sacrifice. In 1027 BCE, the Zhou overthrew the Shang, claiming that Heaven had withdrawn the Shang king's right to rule.

Zhengzhou city

The first Shang capital, Zhengzhou, exhibits many of the economic, social, political, and structural features that typified early cities, but also exemplifies how each state's cities took a different form, related to their own cultural background and requirements. A walled compound—the state's political and religious center—contained royal palaces, built on huge *hangtu* platforms. Royalty and the elite were buried here, with rich grave goods including ritual bronze vessels and jades. Sacrificed dogs and human skulls accompanied religious structures. The massive wall, at least 33ft (10m) high and 120ft (36m) wide at its base, took more than 4 million man-days to build—a visible symbol of the ruler's power to mobilize labor for public works. It imposingly marked the boundary between the court and common people. The output of the many workshops in the surrounding suburban area was largely for the elite. Associated houses show that some artisans enjoyed a higher status than others. The lowest in society, war captives, were not only sacrficed but, as one bone workshop shows, their skulls were used to make bowls.

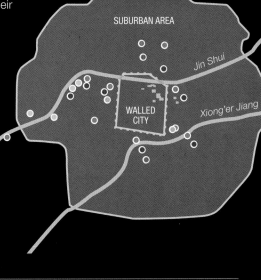

SUBURBAN AREA

Jin Shui

WALLED CITY

Xiong'er Jiang

PLAN OF ZHENGZHOU
Built around 1650 BCE, Zhengzhou comprised a roughly rectangular 830 acre (335 ha) walled city and a 10 sqare mile (25 sq km) suburban area. It was strategically situated on two rivers, which were important communication routes.

KEY

◉ Distillery site	◉ Bone workshop
● Pottery workshop	■ Buildings
○ Burial site	▣ Urban area
◉ Bronze hoard	┅ City wall
◉ Bronze foundry	

Neighbors

Knowledge of the Shang from written sources created the impression that they were the only state in China at this time. However, in the Yangzi River basin, where wet-rice agriculture had supported the growth of well-developed craft industries since the 4th millennium BCE, discoveries in 1986 revealed evidence of a technologically sophisticated urban culture very different in character from the Shang. At Sanxingdui, southwest of the Shang region, a walled city suggests the existence of a major state—perhaps the state of Shu mentioned in Shang oracle bone texts. Housing and workshops manufacturing bronzes, lacquerware, pottery, and jade objects were scattered over an area of 6 sqare miles (15 sq km). At its heart was an area of 1,110 acres (450 ha) surrounded by walls up to 155 ft (47 m) wide. This contained at least one palace and a number of extraordinary ritual pits. These had been carefully filled with objects, very different in style from those of the Shang, made of bronze, gold, jade, stone, turquoise, and ivory, as well as elephant tusks, pottery, cowrie shells, and layers of burned animal bones. The bronzes included human heads, grotesque masks, and a larger than lifesize human figure. A bronze tree nearly 13 ft (4 m) high had flowers, fruit, and birds. Farther west, the cities of Panlongcheng and Wucheng, and the necropolis at Xin'gan, may have been part of the same state.

BRONZE COOKING VESSEL
Sophisticated, reusable multipiece molds were developed in China by 1700 BCE to cast complex shapes like this tripod vessel (*ding*), one of more than 400 bronze vessels from Fu Hao's grave.

BRONZE HEAD
Many bronze heads, including two with masks of gold leaf across their faces, were among the extraordinary objects deposited in the offering pits at Sanxingdui.

American civilizations

Characteristic features of the Olmec in Mesoamerica and the Chavín in the Andes—the first American civilizations—set the enduring pattern in their regions. In each, the shared culture had a religious basis, which was reflected in art and symbolism.

VOTIVE CACHE
Serpentine and jade figurines and celts (axes) arranged as for a ceremony were buried at La Venta, but marked so they could be uncovered when necessary.

The Olmec

Distinctive religious themes characterize Olmec art, and its spread reflects wide acceptance of its underlying beliefs. Sculptures, figurines, rock reliefs, and paintings depict supernatural beings with cleft heads, such as the half-baby half-feline "were-jaguar." Helmets identify the colossal Olmec stone heads—up to 11ft (3.4m) high—as players in a ritual-laden ballgame. This and many other Olmec features, including warfare and sacrifice, a complex calendar, and personal bloodletting, became central to later Mesoamerican culture. Three ceremonial centers span the Olmec period in the Gulf Coast heartland: San Lorenzo, 1200–900 BCE; La Venta, 900–400 BCE; and Tres Zapotes, from 400 BCE when the Olmec were in decline. Dense local settlement suggests ceremonial centers were both religious sites and cities. They feature complex arrangements of stone heads, water channels, massive rectangular thrones, and courtyards; La Venta also had mounds, votive caches, and mosaic pavements depicting jaguar masks.

OLMEC STONE HEAD
Massive basalt heads were erected only in the Olmec heartland. They were probably portraits of Olmec leaders.

OLMEC CIVILIZATION
Olmec art and religious beliefs may have been spread by trade. Highland cultures received ritually significant or prestige materials such as stingray spines and feathers in return for obsidian, jade, and iron ore for making mirrors.

KEY
- ● Olmec and Olmec-influenced sites
- ◼ Olmec heartland
- ◼ Area of Olmec influence

Chavin culture

Rich fisheries supported settled communities on Peru's coast, with some agriculture in inland valleys. Irrigation allowed farming to grow in importance, developing by 2600 BCE in the Supe Valley, where a 125-acre (50ha) town with six temple complexes was established at Caral. Huts and masonry houses indicate a social hierarchy here. From around 2000 BCE, irrigation was more widespread,

CARAL TEMPLE
This mound complex in Caral comprises a circular sunken court and a platform with atrium, a form that was typical of the coastal region.

weaving began, and pottery was widely adopted. Regional styles of architecture and religious art appeared, with U-shaped ceremonial centers on the coast.

The Chavín art style appeared in both highland and coastal sites from around 1200 BCE. It achieved its finest flowering at Chavín de Huantar around 400 BCE although many other impressive contemporary shrines showed strong Chavín influence. Chavín de Huantar was a major ceremonial center, founded around 800 BCE. Its two U-shaped temples had sunken courtyards decorated with large heads and carvings of humans, Amazonian predators, and composite deities (see right). Some of the heads depict the transformation, using hallucinogenic snuff, of a shaman into a jaguar, harpy eagle, or monkey. In the temple's inner sanctum, the Lanzón (see p.229) bore an image of the Chavín supreme deity, the Staff god, a terrifying figure with claws and fangs, a feline face, and snakes for eyebrows and hair. Chavín art also appears on pottery, cloth, and sheet gold.

KEY
- ● Initial period (pre-Chavín) sites
- ● Early Horizon (Chavín and Chavín-influenced) sites
- ● Initial period and Early Horizon sites
- ◼ Chavín heartland
- ◼ Area of Chavín influence

CHAVIN CIVILIZATION
Chavín de Huantar lay strategically at the intersection of trade routes running north–south through the Andes and east–west from the coast to the Amazon Basin. The Chavín cult was widespread in the Andean region, especially around 400–200 BCE.

JAGUAR RELIEF SCULPTURE
The carvings on the walls of the courtyards at Chavín de Huantar depict supernatural beings that are composites of animals and humans. This deity has a jaguar's face and fangs and an eagle's claws.

GLOSSARY

Terms highlighted in **bold** type have their own separate entry.

A

absolute dating
Finding the age of an object—how long ago it first existed or was made—based on the analysis of its composition and in reference to a specific, not relative, time scale. *See also* radiocarbon dating, radiometric dating, relative dating.

Acheulean
The Acheulean is a stone tool culture that was used by **hominins** during the **Stone Age**—from around 1.65 million to 100,000 years ago. It is characterized by handaxes that were worked symmetrically and on both sides. These represent a significant development in technique and sophistication from the **Oldowan**. Acheulean tools are also referred to as "Mode Two" tools, with the earlier Oldowan being known as "Mode One."

anthropoid
A term for the "higher primates," or monkeys and apes (including humans). It contrasts with the "lower primates," or tarsiers, lemurs, lorises, pottos, and galagos (bushbabies).

archaic
An early, "primitive" stage in the evolution of a group of organisms, retaining some features of the predecessors or ancestors.

artifact
An object made or adapted by humans, such as a stone tool or a pottery vase, rather than an object already existing in nature.

B

Beaker culture
Groups of people who typically produced a distinctive style of beaker or pottery vessel with an upside-down bell shape, and other associated cultural artifacts, mainly in Western Europe from about 4,800 to 3,800 years ago.

biochronology
The use of fossils to establish a timescale of relative ages that can be applied over wide areas, even worldwide, by means of correlation (comparison) of fossil assemblages and the enclosing rocks.

bipedalism
Walking habitually or routinely on two legs, as in humans and certain other mammals, and birds.

braincase
The domed casing of the upper skull bones that forms a protective covering for the brain. Sometimes used with the same meaning as "cranium," as in cranial capacity or brain size. *See also* cranium.

Bronze Age
A major stage in human cultural development between the **Stone Age** and **Iron Age**, when the alloy (combination) of copper and tin, known as bronze, was used for tools as well as ornaments. It appeared at different times in different regions from about 5,400 years ago, but was absent in some areas.

C

catarrhine
A member of the Old World monkey and ape group (based in Europe, Africa, and Asia). *See also* platyrrhine.

Chromosome
A coiled, long, threadlike molecule of genetic material—deoxyribonucleic acid (**DNA**). Each kind of animal or plant has a characteristic number of chromosomes in each of its cells. For example, living humans have 46 chromosomes, in 23 pairs.

city-state
A region with a shared culture based around a city, including the surrounding towns, villages, and farmlands, and acting independently of other city-states.

clade
A group of species that includes all the evolutionary descendants of a given ancestor—the "common ancestor."

cladistics
A method of grouping and classifying organisms based on units called **clades**. It focuses on shared derived features that have been inherited from a common ancestor, rather than features that have evolved independently.

clavicle
The collar bone, a bone found in the shoulder or pectoral region of mammals, birds, and some reptiles and amphibians.

Copper Age (Chalcolithic)
A stage in human cultural development, between the **Stone Age** and **Bronze Age**, when the copper was worked, mainly for prestige objects, often alongside gold, silver, and lead. It occurred in some areas as long as 7,000 years ago, but not at all in others.

Copper Culture
A term applied to various American cultures known for cold-working the metal copper, from 6,000 years ago onward.

Corded Ware Culture
Groups of people who typically produced a distinctive style of ornamentation or decoration on their pottery, along with other associated cultural artifacts, mainly in Northern and Eastern Europe from about 5,000 to 4,300 years ago.

core
A core is an archaeological artifact created when a basic block of raw stone material has been reduced by the detachment of flakes, usually by a **hammerstone**. Cores were either discarded or used as choppers and handaxes.

cranial
Relating to the **cranium**.

cranium
The upper domed part of the skull that encases and protects the brain. *See also* braincase.

D

derived
A derived characteristic or feature is one that evolves from and appears later than the original or ancestral form of the characteristic. *See also* cladistics.

DNA
Deoxyribonucleic acid is a very long molecule made up of small individual units. DNA is found in the

cells of all animals and other living things; the order of the small units "spells out" the genetic instructions (genes) of the animal. *See also* chromosome, gene, genome.

E

Eocene
The second epoch of the Paleogene period, from 56 to 34 million years ago.

era
A unit of the geological timescale that is a division of an eon and is itself divided into periods. We live in the Cenozoic era, which began 65 million years ago.

F

femur
The thigh bone, linking the pelvis or hip bone at the hip joint to the lower leg at the knee.

foramen magnum
The large opening at the base of the skull, through which the spinal cord—the body's main nerve—passes and connects to the brain above.

G

genes
Units of inheritance consisting of certain lengths or segments on a molecule of **DNA**, carrying information that contributes to a particular feature of a living thing, for example, eye color, building enamel for teeth, or making hemoglobin for blood.

genome
All of an organism's genetic or hereditary information, as represented in its full set of genes.

gracile
Slender, slim, or lightly built. *See also* robust.

great ape
A member of the group that includes living humans, chimpanzees, gorillas, orangutans, and their ancestors.

H

hammerstone
A large stone or cobble used to strike flakes or pieces from the **core** (stone) being worked.

haplorrhine
A member of the primate group that includes tarsiers, monkeys, and apes (including humans).

Holocene
The most recent epoch of the Neogene period, lasting from 11,700 years ago to the present. It is also designated as part of the Quaternary period.

hominid
A member of the biological family *Hominidae* (**great apes**), including humans, chimpanzees, gorillas, orangutans, and their ancestors. Previously, this term was used to describe just humans and their ancestors, but the new classification better reflects our close relationship with the other great apes.

hominin
Refers specifically to a member of the biological tribe *Hominini*. Some scientists include chimpanzees within this group, whereas others use it to refer to humans and their ancestors, after the split from the last common ancestor of the chimpanzee around 7 million years ago. It is used in the latter sense in this book.

hominoid
A member of the biological superfamily *Hominoidea*, including **hominids** or **great apes** (humans, chimpanzees, gorillas, and orangutans) and the hylobatids or lesser apes (gibbons and siamangs).

humerus
The upper arm or forelimb bone, connecting the shoulder bones to the lower arm at the elbow.

hunter-gatherers
People who hunt wild animals and gather plant foods, rather than using farming methods such as agriculture or domesticating animals.

I

ice age
Any episode in which the Earth's surface temperatures were much lower than today and the ice cover was more extensive. The term "Ice Age" is often applied to the most recent colder period that peaked at the **last glacial maximum**.

Iron Age
A major stage in human cultural development that followed the **Bronze Age**, when most metal tools and weapons were manufactured from iron and its alloys. It appeared at different times in different regions, but became widespread in the Old World during the 1st millennium BCE.

K

knapping
Striking a piece of stone to achieve a desired shape, such as using a **hammerstone** to remove small flakes.

knuckle-walking
A method of locomotion using all four limbs, with the feet flat on their soles, but the fingers of the hands flexed and the front of the body supported on the middle finger bones.

L

Last Glacial Maximum (LGM)
Refers to the period of the maximum extent of the ice sheets during the last glacial episode, between 26,500 and 18,000 years ago. The term "post-glacial" refers to the period when the ice sheets were retreating.

Lomekwian
The Lomekwian is the earliest known stone tool culture. It was used by **hominins** in the early **Stone Age**—around 3.3 million years ago. It is characterized by very large **core** stones and flakes that were crudely worked. The Lomekwian predates the **Oldowan** by about 700,000 years.

M

mandible
The lower jaw bone, which is hinged to the rest of the skull at the jaw joints.

maxilla
The upper jaw bone, which in most mammals is fixed firmly to the rest of the skull.

megalith
A large stone or rock, often weighing many tons, that has been used to construct buildings or monuments, either alone or with other stonework. *See also* microlith, passage grave.

Mesoamerica
The geographical region connecting

North and South America, usually from present-day Central Mexico to Costa Rica/Panama. Mesoamerican is a term for cultures that have occurred in this area.

Mesolithic

Middle Stone Age, the period between the **Paleolithic** (Old Stone Age) and **Neolithic** (New Stone Age). It lasted for different lengths of time in different regions, from about 11,500 years ago and refers to the **hunter-gatherer** societies of the period after the last **Ice Age**. In Africa, the term "middle Stone Age" has a different meaning, signifying the middle stage of the Paleolithic period. *See also* Stone Age.

Mesopotamia

Meaning "between rivers," Mesopotamia was the West Asian region between the Tigris and Euphrates rivers, which corresponds to today's southeast Turkey, northeast Syria, Iraq, and southwest Iran.

microlith

A very small stone blade, used as a tool or weapon, usually fashioned by striking, shaping, and flaking. Characteristic of the **Mesolithic** period.

Minoan civilization

A civilization based on the Mediterranean island of Crete, beginning around 4,000 years ago and collapsing around 3,450 years ago. Minoans had **Bronze Age** technologies, advanced agriculture, palace economies, and a flourishing trading network within the eastern Mediterranean using sea-going merchant ships.

Miocene

The first epoch of the Neogene period, from 23 to 5.3 million years ago.

mitochondrial DNA (mtDNA)

Small amounts of the genetic material **DNA** located in the parts of cells called mitochondria, rather than nuclear DNA, which is found in the nucleus and forms the vast bulk of a cell's DNA. mtDNA is inherited through the female line and changes or mutates relatively rapidly, and so is a useful resource for tracing patterns of heredity.

molecular clock

A method of aging or dating events using the changes or mutations in molecules. For example, it employs average or typical mutation rates to compare two similar molecules and work out how long ago they diverged. It can be done especially with lengths of **DNA**, or working back from proteins such as hemoglobin found in blood, which are made according to genetic instructions that mutate.

morphology

The forms and shapes of living things, including their external and internal structures and features, as well as colors and patterns.

mutation

A change in the structure or sequence of genetic material, usually **DNA**. Such changes may have a positive or negative effect on the organism or no effect at all. Mutations may be caused by "natural" copying errors or to outside agents such as radiation, viruses, or certain chemicals called mutagens.

mya

Abbreviation for millions of years ago or million years ago.

Mycenaean civilization

A civilization based in southern Greece, beginning more than 3,600 years ago and fading by 3,100 years ago. Mycenaeans had **Bronze Age** technologies, fortified palaces, a warrior society, and a widespread and varied trading network in the Mediterranean.

N

natural selection

An important part of evolution where some organisms with certain inherited features or characteristics are better suited, or adapted, to their environment, and so survive and pass on their features to the next generation. Those less well adapted do not survive.

Neolithic

New Stone Age, the period between the **Mesolithic** (Middle Stone Age), and usually the following **Copper**, **Bronze**, or **Iron Ages**. It appeared at different times in different regions from about 11,500 years ago and saw the development of farming and a settled way of life, with domesticated crops and livestock. *See also* Stone Age.

O

Oldowan

The Oldowan is the earliest stone tool culture. It was used by **hominins** in the early **Stone Age**—from around 2.5–1.5 million years ago. It is characterized by crudely worked **core** stones and simple flakes. Oldowan tools are also referred to as "Mode One" tools, with the later **Acheulean** being known as "Mode Two."

Oligocene

The third epoch of the Paleogene period, from 34 to 23 million years ago.

P

paleoanthropology

The study of ancient or prehistoric humans and how they lived, using evidence such as body fossils, trace fossils like footprints and tooth marks, signs of dwellings, artifacts such as stone tools, clothing, and other clues.

Paleocene

The first epoch of the Paleogene period, from 65 to 56 million years ago.

Paleolithic

Old Stone Age, the first period of the major phase in human cultural development known as the **Stone Age**. It began with the earliest simple stone tools and other artifacts in Africa more than two million years ago.

passage grave

A **Neolithic** grave, tomb chamber, or burial site incorporating a passageway that is usually quite narrow and made of large stones or **megaliths**. Passage graves are often covered with a mound of earth or other materials.

pelvis

The hip bone, the large flange of bone at the hip joint, linking the lower backbone to the femur or thigh bone of the hind limb.

phytoliths

Microscopic, hard, stonelike parts of plants such as grasses, palms, and baobabs, which usually confer support and rigidity. Their shapes vary according to the plant type. Phytoliths are very tough and resistant, so they form valuable archaeological evidence for plant growth and use by ancient peoples.

platyrrhine

A member of the New World monkey group (based in the Americas).

Pleistocene

The first epoch of the Quaternary period, from 2.6 million to about 12,000 years ago. Geologists have only recently extended the beginning of the Pleistocene back from 1.8 million years ago.

Pliocene

The second epoch of the Neogene period, from 5.3 to 2.6 million years ago.

Pre-Pottery Neolithic (PPN)

A localized name in Israel, Palestine, and Syria, for the Aceramic ("without pottery") stage of the **Neolithic** period in West Asia, lasting from about 10,600 to 7,500 years ago.

primates

The biological order containing **haplorrhines** and **strepsirrhines**.

R

radiocarbon dating

A method for finding the actual age, in thousands of years, for objects containing carbon—which include all living things. It is a form of **absolute dating** that measures the breakdown or decay of radioactive forms of carbon. In practice it is limited to around the last 60,000 years.

radiometric dating

Finding the age of an object—how long ago it first existed or was made—by measuring the proportions of radioactive substances it contains, and working backward using their known breakdown or decay rates. *See also* absolute dating.

relative dating

Finding the age of an object—how long ago it first existed or was made—by comparing it (relating it) to objects found with or near it, such as fossils of known ages in the rock layers above or below it. *See also* absolute dating.

remote sensing

The imaging of an object or area from a distance, without coming in direct contact with it, mainly through the use of aerial photography or satellite imagery or by measuring reflected and emitted radiation.

robust

Sturdy, thick-set or strongly built. *See also* gracile.

S

sagittal crest

A ridge of bone running from front to back along the middle of the top of the skull, typically seen in the living mature male gorilla, and in sexually mature male orangutans. It also occurs in various fossil apes including some hominins. It usually indicates the anchor site of powerful jaw muscles for prolonged chewing.

scapula

The shoulder blade, the large flange of bone at the shoulder joint, linking the upper backbone of the torso to the humerus bone of the upper forelimb.

sedentism

A way of life that marks the transition from a wandering, nomadic lifestyle to staying permanently in the same place.

sexual dimorphism

The degree of physical difference between adult males and females of the same species, in addition to the sex organs. This may be seen in body size, shape, appearance, and structure.

speciation

The formation of new species, for example, when groups of the same species become isolated on different islands, adapt to the new conditions on their islands, and evolve to become distinct species.

Stone Age

A major stage in human cultural development involving the use and modification of stones as tools, weapons, and other artifacts, before the development of metalworking. It is usually divided into three phases. *See* Paleolithic, Mesolithic, and Neolithic.

strepsirrhine

A member of the primate group including lemurs, lorises, pottos, and galagos (bushbabies).

suture

In the skeleton, a joint where two bones are firmly linked together and cannot move in relation to each other. Several bones of the skull are joined at sutures, which can sometimes be seen as wavy lines.

T

tell

A large hump or mound, usually relatively flat-topped, formed by the accumulation of buildings, construction materials, utensils, garbage, and the general debris of living from a long period of human settlement, or perhaps several periods with abandoned times between.

tumulus (plural tumuli)

A pile or mound of earth, stones, and other materials built over a tomb, grave, or burial site. Tumuli are sometimes called burial mounds, kurgans, or barrows.

type specimen

The "original" specimen (or a group of them) from which a particular species or other classification group is first defined, described, and officially named in the scientific literature.

Z

ziggurat

A step-pyramid—a structure with several layers, usually of stone, becoming smaller with height to create a series of steps. The term "ziggurat" is sometimes applied only to these structures in **Mesopotamia** and nearby areas, while others use it for all similar structures, including those in Egypt and **Mesoamerica**.

INDEX

Page numbers in **bold** refer to main entries; those in *italics* refer to captions to illustrations.

ACKNOWLEDGMENTS

For the revised edition, Dorling Kindersley would like to thank Dharini Ganesh and Riji Raju for editorial assistance; Sanjay Chauhan and Yashasvi Choudhary for design assistance; Taiyaba Khatoon and Sakshi Saluja for picture research assistance; Dheeraj Singh and Syed Mohammad Farhan for DTP assistance; and Juhi Sheth for jacket design assistance.

The publisher would like to thank the following for their kind permission to reproduce their photographs:

Genetic Tree on p.178 redrawn from fig. 0.3, *Out of Eden*, Oppenheimer 2003; arrows on map on pp.176–77 in part based on map in *Out of Eden*, 2003, and fig. 1 of *Quaternary International* 2009.

(Key: a-above; b-below/bottom; c-center; f-far; l-left; r-right; t-top)

6-7 Alamy Stock Photo: Horst Klemm / Greatstock. 7 Dave Stevens. 10-243 Corbis: Bill Ross (Background). **11 Dorling Kindersley:** NASA / Digitaleye / Jamie Marshall (br); University Museum of Zoology, Cambridge (ca). **12 Corbis:** Martin Rietze / Westend61 (bc); Mike Theiss / Ultimate Chase (tc). **13 Corbis:** Franck Guiziou / Hemis (b). **Science Photo Library:** Mark Pilkington / Geological Survey of Canada (tr). **14 Corbis:** Rainer Hackenberg. **Wikipedia, The Free Encyclopedia:** from The Geoscientist v 18, n 11, portrait by Hugues Fourau (br). **15 Dorling Kindersley:** University Museum of Zoology, Cambridge (cr). **16 The Natural History Museum,**

London: (bl, bc). **16-17 Dorling Kindersley:** Natural History Museum, London. **17 Corbis:** James L Amos (cl); George HH Huey (tr). Science Photo Library: Professor Matthew Bennett, Bournemouth University (bl). **18 Dorling Kindersley:** Natural History Museum, London (bl). SeaPics.com: Mark V. Erdmann (cl). **20 Getty Images. 21 Corbis:** Nigel Pavitt / JAI (tr); Vienna Report Agency (br). **Dorling Kindersley:** National Museum of Copenhagen (bc). **Getty Images:** Kenneth Garrett / National Geographic (tl). **22 Science Photo Library:** John Reader (bl); Javier Trueba / MSF (t). **23 Alamy Images:** John Elk III (bl). **Corbis:** Visuals Unlimited (br). **Science Photo Library:** Pascal Goetgheluck (ca); Pasquale Sorrentino (t); David Scharf (cr). **24 Science Photo Library:** Pascal Goetgheluck (cl); Philippe Psaila (tl). **24-25 Corbis:** Arne Hodalic (t). **Science Photo Library:** Pasquale Sorrentino (b). **25 Corbis:** Reuters / Ho (tc). **Science Photo Library:** Volker Steger (br). **26 Corbis:** Christophe Boisvieux (bc). **Getty Images:** AFP (bl). **Hull York Medical School:** Centre for Anatomical and Human Sciences. Produced by Dr. Laura Fitton with support from BBSRC (grant BB / E013805 / 1) to Professors Paul O'Higgins (HYMS, University of York) and Michael Fagan (Dept. Engineering, University of Hull) (br). **Science Photo Library:** Javier Trueba / MSF (cl). **26-27 Science Photo Library:** Javier Trueba / MSF. **27 Science Photo Library:** D Roberts (br). **Professor Tanya M. Smith:** (bc). **28 National Institute of Standards**

and Technology / NIST: David S. Strait / University of Albany, SUNY, Photograph of Sts 5 by Gerhard Weber (bc). **Science Photo Library:** Philippe Plailly (bl). **32 Dorling Kindersley:** Natural History Museum, London (ca). **33 Corbis:** Hemis / Jean-Daniel Sudres (b). **Getty Images:** AFP (cra). **The Natural History Museum, London:** (tr). **36-37 Corbis:** Specialist Stock. **37 Corbis:** Wayne Lawler / Ecoscene (cr). **41 Dr. Doug M. Boyer:** (tr). **Jens L. Franzen:** Gingerich PD, Habersetzer J, Hurum JH, von Koenigswald W, et al. 2009 Complete Primate Skeleton from the Middle Eocene of Messel in Germany: Morphology and Paleobiology. PLoS ONE 4(5): e5723. doi:10.1371 / journal. pone.0005723 (br). **42-43 Corbis:** Frans Lanting (c). **42 Corbis:** Frans Lanting (cla). **43 Dorling Kindersley:** Harry Taylor, Courtesy of the Natural History Museum, London (cr). **National Science Foundation, USA:** Erik Seiffert, Stony Brook University (br). **44 Dorling Kindersley:** Thomas Marent (cl). **Photographs courtesy of Andrea L. Jones:** (bl, bc). **45 Alamy Images:** Terry Whittaker (bl). **Corbis:** (c); Thomas Marent / Visuals Unlimited (tr); Kevin Schafer (bc). **Dorling Kindersley:** Rough Guides (br). **Getty Images:** Photo 24 / Brand X Pictures (cr). **46 Dorling Kindersley:** Natural History Museum, London (tr, cb). **National Museums of Kenya:** (fbl, bl). **47 Corbis:** Theo Allofs (tr). **Dorling Kindersley:** Jamie Marshall (cra); Rough Guides (tc). **48-49 Corbis:** Frank Lukassek. **50 Corbis:** Frans Lanting (bl). **Dorling Kindersley:** Oxford University Museum of Natural History (tr);

Rough Guides (tl). **51 Corbis:** Petr Josek / Reuters. **52 © Bone Clones, www.boneclones.com:** (tl, tc, cl, c, bc). **53 © Bone Clones, www.boneclones.com:** (cl, tc). **54 FLPA:** Albert Lleal / Minden Pictures (bl). **naturepl.com:** Anup Shah (t). **55 Corbis:** Bettmann (crb); Frans Lanting (cr); DLILLC (tc). **Getty Images:** John Moore (cla). **Science Photo Library:** Tom McHugh (ca). **64 Professor Michel Brunet:** (br). **Dorling Kindersley:** Gary Ombler (bc, bl). **Getty Images:** Alain Beauvilain / AFP (tr). **65 Alamy Images:** WorldFoto (t). **68 Alamy Images:** Mike Abrahams (bc). **Camera Press:** Marc Deville / Gamma (tr). **69 Corbis:** Markus Botzek (cr). **70 Alamy Images:** FLPA (br); F. Scholz / Arco Images GmbH (crb). **Los Alamos National Laboratory (LANL):** (cl). **71 Corbis:** Reuters / Science / AAAS (cr, tr, bc). **Reuters:** T. White / Science / AAAS (l). **74 Kenneth Garrett:** (b). **National Museums of Kenya:** (tr). **75 Alamy Images:** Kolvenbach (b). **Professor Michel Brunet:** (t). **78 Getty Images:** AFP (bc). **Science Photo Library:** John Reader (tl). **79 PNAS:** Yohannes Haile-Selassie, Bruce M. Latimer, Mulugeta Alene, Alan L. Deino, Luis Gibert, Stephanie M. Melillo, Beverly Z. Saylor, Gary R. Scott, and C. Owen Lovejoy.; An early Australopithecus afarensis postcranium from Woranso-Mille, Ethiopia. doi: 10.1073 / pnas.1004527107 ; PNAS July 6, 2010 vol. 107 no. 27 12121-12126 (br). **80 Science Photo Library:** John Reader (tl). **Paul Szpak:** (tr). **80-81 Science Photo Library:** John Reader (b). **81 Dikika Research Project:** photo by Curtis Marean; Christoph P. E. Zollikofer et al,

Virtual cranial reconstruction of Sahelanthropus tchadensis, Nature 434, 755-759 (7 April 2005), (doi:10.1038 / nature03397), figure 4, reprinted by permission from Macmillan Publishers Ltd (tr). **Raichlen DA,:** Gordon AD, Harcourt-Smith WEH, Foster AD, Haas WR Jr, 2010 Laetoli Footprints Preserve Earliest Direct Evidence of Human-Like Bipedal Biomechanics. PLoS ONE 5(3): e9769. doi:10.1371 / journal.pone.0009769 (tl). **88 Alamy Images:** AfriPics.com (b). **Science Photo Library:** John Reader (t). **89 Dorling Kindersley:** Gary Ombler (c, cr); Harry Taylor, Courtesy of the Natural History Museum, London (bc). **Kenneth Garrett:** (br). **Science Photo Library:** John Reader (bl). **92 Alamy Images:** Hemis (tr). **Dorling Kindersley:** Gary Ombler (bc). **93 Science Photo Library:** John Reader (tc, tr). **94 Corbis:** Brian A Vikander (b). **Science Photo Library:** Des Bartlett (tr). **95 PLOS ONE:** © 2008 Ungar et al./Ungar PS, Grine FE, Teaford MF (2008) Dental Microwear and Diet of the Plio-Pleistocene Hominin Paranthropus boisei. PLoS ONE3(4): e2044. https://doi.org/10.1371/journal.pone.0002044 (clb). **95 Dorling Kindersley:** Natural History Museum, London (tl). **The Natural History Museum, London:** (c, br). **100 Science Photo Library:** John Reader (tr); John Reader (bc). **101 Dorling Kindersley:** Natural History Museum, London (br, tr). **102 Dorling Kindersley:** Natural History Museum, London (cra). **Getty Images:** National Geographic / David S. Boyer (c). **Science Photo Library:** John Reader (tl). **103 Alamy Images:** David Keith Jones / Images of Africa Photobank (tr); Paul Maguire (b). **106 Kenneth Garrett:** (t). **110 Alamy Images:** Danita Delimont / Kenneth Garrett (br). **Kenneth Garrett:** (t). **111 Copyright Clearance Center - Rightslink:** Nature 449, 305-310 (20 September 2007) | doi:10.1038 / nature06134, David Lordkipanidze et al, Postcranial evidence from early Homo from Dmanisi, Georgia, figure 2 from, reprinted by permission from Macmillan Publishers Ltd (cl). **Kenneth Garrett:** (tr). **Getty Images:** National Geographic / Kenneth Garrett (bl). **116 Alamy Images:** Marion Kaplan (tr); John Warburton-Lee Photography (cl). **Copyright Clearance Center - Rightslink:** Science 27 February 2009: 323 (5918), 1197-1201 figure 3, Matthew R. Bennett, et al, Early Hominin Foot Morphology Based on 1.5-Million-Year-Old Footprints from Ileret, Kenya © 2009 The American Association for the Advancement of Science (bc). **117 Dorling Kindersley:** Trish Gant / National Museum of Kenya (l); Gary Ombler (tc). **118 Dorling Kindersley:** Dave King / Courtesy of the Pitt Rivers Museum, University of Oxford (tr); The Natural History Museum (tc). **Faisal A:** Stout D, Apel J, Bradley B (2010) The Manipulative Complexity of Lower Paleolithic Stone Toolmaking. PLoS ONE 5(11): e13718. doi:10.1371 / journal.pone.0013718 (br). **Science Photo Library:** John Reader (c). **119 Alamy Images:** Karin Duthie (br). **Dorling Kindersley:** Nigel Hicks (t). **124 Alamy Images:** (b). **Getty Images:** AFP (cl). **Copyright NNM, Leiden, The Netherlands:** (tr). **Science Photo Library:** John Reader (c). **125 Alamy Images:** Natural History Museum, London (bl). **Archives of American Art, Smithsonian Institution:** photo courtesy Human Origins Program / Science 3 March 2000: 287 (5458), 1622-1626, figure 2. Mid-Pleistocene Acheulean-like Stone Technology of the Bose Basin, South China Hou Yamei, Richard Potts, Yuan Baoyin, Guo Zhengtang, Alan Deino, Wang Wei, Jennifer Clark, Xie Guangmao and Huang Weiwen, (c) 2000. **The Natural History Museum, London:** (cl). **130 Science Photo Library:** Javier Trueba / MSF (cl, cr, br, bc, cb). **131 Getty Images:** AFP (bc). **Science Photo Library:** Javier Trueba / MSF (skeletal bones, br). **136 Getty Images:** AFP (cl). **Science Photo Library:** (cr); John Reader (tr). **137 The Trustees of the Natural History Museum, London:** (br). **137 The Natural History Museum, London:** (cr, crb). **142 Alamy Images:** Banana Pancake (c). **Karen L. Baab, Stony Brook University:** photo courtesy of Peter Brown, University of New England, Australia (br). **Dorling Kindersley:** Gary Ombler (tr, cra). **143 Adam Brumm, University of Wollongong:** (c). **Kenneth Garrett:** (br). **Getty Images:** Kenneth Garrett / National Geographic (bc). **W.L. Jungers:** (l). **144-145 Flickr.com:** Rosino (http://www.flickr.com <http//www.flickr.com>/photos/84301190@N00/1525434007/). **150 Kenneth Garrett:** (br, cl). **151 © Bone Clones, www.boneclones.com:** (l). **152 © Bone Clones, www.boneclones.com:** (t). **Dorling Kindersley:** Natural History Museum, London (bl). **PNAS:** João Zilhãoa, Diego E. Angelucci, Ernestina Badal-García, Francesco d'Errico, Floréal Daniel, Laure Dayet, Katerina Douka, Thomas F. G. Higham, María José Martínez-Sánchez, Ricardo Montes-Bernárdez, Sonia Murcia-Mascarós, Carmen Pérez-Sirvent, Clodoaldo Roldán-García, Marian Vanhaeren, Valentín Villaverde, Rachel Wood, & Josefina Zapata. Symbolic use of marine shells and mineral pigments by Iberian Neandertals. doi: 10.1073 / pnas.0914088107 ; PNAS January 19, 2010 vol. 107 no. 3 1023-1028 ; (bc). **153 Corbis:** Anup Shah (tr). **Science Photo Library:** Javier Trueba / MSF (bl). **158 Getty Images:** Barcroft Media (cla). **National Geographic Creative:** Robert Clark (cra, b). **159 National Geographic Creative:** Robert Clark (l); Stefan Fichtel (tc). **164 Alamy Images:** Hemis (cr); Ariadne Van Zandbergen (b). **Getty Images:** National Geographic (tc). **The Natural History Museum, London:** (tr, c). **165 The Natural History Museum, London:** (tl). **166 Marlize Lombard:** and Laurel Phillipson, Indications of bow and stone-tipped arrow use 64 000 years ago in KwaZulu-Natal, South Africa, Antiquity 2010, Vol: 84 No: 325 pp635–648 (http: / / www.antiquity.ac.uk / Ant / 084 / ant0840635.htm), photo courtesy Marlize Lombard (tr). **Science Photo Library:** John Reader (ca, bl, cl). **167 Science Photo Library:** Pascal Goetgheluck. **177 Getty Images:** Peter Adams / The Image Bank (crb); Peter Adams / Photographer's Choice (fbr); Matthias Clamer / Stone+ (br); Brad Wilson / Stone (fcrb). **178 Science Photo Library:** Don Fawcett (tr). **179 Science Photo Library:** Eye of Science (br); James King-Holmes (bl). **180 Alamy Images:** Ancient Art & Architecture Collection Ltd (bc); FLPA (t); Danita Delimont / Kenneth Garrett (c). **Corbis:** Visuals Unlimited / Carolina Biological (br). **KH Wellmann, Frankfurt am Main:** (bl). **181 Alamy Images:** Danita Delimont / Kenneth Garrett (bl). **Getty Images:** AFP (tl). **Science Photo Library:** John Reader (cr). **182 Niedersächsisches Landesamt für Denkmalpflege (NLD):** Peter Pfarr (bl). **Science Photo Library:** Javier Trueba / MSF (c). **182-183 Getty Images:** AFP. **183 Kenneth Garrett:** (tr). **184 Alamy Images:** Kenneth Garrett / Danita Delimont (cr). **Corbis:** Eberhard Hummel (crb); Hanan Isachar (bl). **University of Bergen, Norway:** Christopher Henshilwood & Francesco d'Errico (tl). **185 Science Photo Library:** John Reader. **186 Corbis:** EPA (c); NASA (bl); Charles & Josette Lenars (br). **Dr. Alice Roberts:** (tc). **187 Corbis:** fstop / Marc Volk (t). **Kenneth Garrett:** (cr). **Colin Groves, Australian National University:** published by permission of Traditional Owners (bl). **Traditional Owners, published by permission of. 188 Dr. Alice Roberts**. **189 akg-images:** (cr). **Getty Images:** AFP (br). **Mircea O. Gherase:** (bl). **190 The Natural History Museum, London:** (bl). **José Paulo Ruas:** courtesy João Zilhão, ICREA Research Professor, Universitat de Barcelona (http: / / www.bristol.ac.uk / news / 2010 / 6777.html) (r). **191 Corbis:** DPA (tl). **Kenneth Garrett:** (bl, br). **Max Planck Institute for Evolutionary Anthropology:** Bence Viola (tr). **192-193 Alamy Images:** Arctic Images (t). **192 Dr. Alice Roberts:** courtesy of the Hermitage Collection (bl). **Science Photo Library:** Ria Novosti (br). **193 The Natural History Museum, London:** (br). **194 Alamy Images:** Phil Degginger (br). **Corbis:** Reuters (cl). **Kenneth Garrett:** (bc, bl). **195 Kenneth Garrett:** (tl, b). **196-197 Getty Images:** Altrendo (b). **196 Professor Glenn Summerhayes, Anthropology, University of Otago, Dunedin:** (cr). **The University Of Auckland:** Courtesy of the Anthropology Photographic Archive, Department of Anthropology (c). **197 Alamy Images:** Bill Brooks (tl). **200 Corbis:** AlaskaStock (main image). **201 Alamy Images:** Graham Barclay (cra); Paul Felix Photography (clb). **Corbis:** Nadia Isakova / Loop Images (bl). **Norwich Castle Museum and Art Gallery:** (ca). **202 Alamy Images:** Interfoto (cl). **Torben Dehn, Heritage Agency of Denmark:** (b). **Dorling Kindersley:** Chris Gomersall Photography (c). **203 Dr. Uzi Avner:** Journal of Arid Environments, Vol 74, Issue 7, July 2010, 808-817, A. Holzer, U. Avner, N. Porat and L.K. Horwitz, Desert kites in the Negev desert and northeast Sinai: their function, chronology and ecology (c) 2009 with permission from Elsevier, reprinted from (tl). **The Trustees of the British Museum:** (bl). **Corbis:** Peter Johnson (tr). **National Museum Of Denmark:** Lennart Larsen (br). **204 Corbis:** Hoberman Collection (bl). **204-205 Corbis:** Gavin Hellier / JAI (b); Kazuyoshi Nomachi (t). **205 Corbis:** David Muench (tr). **Raveesh Vyas:** (bc). **206 Getty Images:** DEA / G. Dagli Orti (b). **207 Corbis:** Stephanie Maze. **208 Alamy Images:** Gianni Dagli Orti / The Art Archive (cla). **Dorling Kindersley:** Rough Guides (b). **209 Alamy Images:** Interfoto (cla). **© CNRS Photothèque:** Catherine Jarrige (bc). **Corbis:** Nathan Benn / Ottochrome (tr); Buddy Mays (bl). **Science Photo Library:** R. Macchiarelli / Eurelios (crb). **210 Camera Press:** Berthold Steinhilber / laif (cl, br, bl). **211 Alamy Images:** dbimages (t). **Camera Press:** Berthold Steinhilber / laif (bl, br). **212-213 Alamy Images:** Roberto Esposti. **212 Dorling Kindersley:** Museum of London (bl). **NASA:** Image Science and Analysis Laboratory, NASA-Johnson Space Center. "The Gateway to Astronaut Photography of Earth." (clb). **Press Association Images:** AP Photo / Mike Hettwer, National Geographic Society (c). **213 Corbis:** DLILLC (br); Philip Gould (tr). **Photolibrary:** Guenter Fischer / Imagebroker (cra). **214 Dorling Kindersley:** Rough Guides (t). **Brian Ritchie**. **215 Corbis:** Sakamoto Photo Research Laboratory (bc). **Rowan Flad:** (cl). **Getty Images:** De Agostini Picture Library (tc). **C.F.W. Higham:** (br). **216 Corbis:** Reuters / Ho (cl). **216-217 Last Refuge:** Adrian Warren. **217 The Bridgeman Art Library:** Bildarchiv Steffens (br). **218 Alamy Images:** Jim West (bl). **Corbis:** Gianni Dagli Orti / The Picture Desk Limited (bc); Wolfgang Kaehler (c). **Getty Images:** Claudio Santana / AFP (br). **219 Corbis:** Nigel Pavitt / JAI. **220 Alamy Images:** Mark Boulton (tl). **Corbis:** Jose Fuste Raga (b). **Dorling Kindersley:** Courtesy of The Museum of London / Dave King (c). **221 Corbis:** Ren Junchuan / Xinhua Press (b); Charles O'Rear (c). **Dorling Kindersley:** British Museum (tl); Science Museum, London (tr). **222 Corbis:** Les Pickett / Papilio (tl). **Dorling Kindersley:** Ashmolean Museum, Oxford (bl); Peter Hayman (c) The British Museum (c). **Getty Images:** (cl). **222-223 Dorling Kindersley:** Michel Zabe / CONACULTA-INAH-MEX. Authorized reproduction by the Instituto Nacional de Antropologia e Historia.. **223 Alamy Images:** Interfoto (br). **Corbis:** Fritz Polking / Visuals Unlimited (tr). **Getty Images:** E. Papetti / De Agostini Picture Library (tl). **224-225 Alamy Images:** Edwin Baker . **225 Alamy Images:** Liu Xiaofeng / TAO Images Limited (tr). **Corbis:** Jonathan Blair (cl). **Dorling Kindersley:** ARF / TAP (Archaeological Receipts Fund) (br). **226 Alamy Images:** (c). **Dorling Kindersley:** Australian Museum, Sydney (cl); The British Museum (tr, cr). **227 360 Degrees Research Group (www.360derece.info/english/360_eng.htm) :** (br). **Corbis:** (tl). **Dorling Kindersley: The British Museum** (cra); **The Natural History Museum** (tc); **Natural History Museum, London** (bl, bc). **Getty Images: Robert Harding World Imagery** (crb). **228 Corbis:** Amar Grover / JAI (tl); Sandro Vannini (tr). **Dorling Kindersley:** The Trustees of the British Museum (bl). **229 Corbis:** Alfredo Dagli Orti / The Art Archive / The Picture Desk Limited; (tr); Richard List (tl); Michael S. Yamashita (cra). **Getty Images:** Kenneth Garrett / National Geographic (br). **230-231 Corbis:** Chris Hill / National Geographic Society (t). **230 Michael Fox:** (bc). **231 Corbis:** Gianni Dagli Orti (br); Adam Woolfitt (bc). **232 Corbis:** Hervé Collart (b). **Robert Harding Picture Library:** age fotostock (t). **233 Dorling Kindersley:** Judith Miller / Helios Gallery (bl); National Museum, New Delhi (tr); Courtesy of the University Museum of Archaeology and Anthropology, Cambridge / Dave King (cr). **Robert Harding Picture Library:** Richard Ashworth (cl). **234 Corbis:** Nik Wheeler (clb). **© Richard H. Meadow / Harappa.com, courtesy Dept. of Archaeology and Museums, Govt. of Pakistan:** (tr). **235 Dorling Kindersley:** The Trustees of the British Museum (br); National Museum, New Delhi (tr). **236-237 Dorling Kindersley:** The Trustees of the British Museum. **237 Dorling Kindersley:** The Trustees of the British Museum (b). **238 Alamy Images:** Ancient Art & Architecture Collection Ltd (tl). **239 Alamy Images:** Ancient Art & Architecture Collection Ltd (tl). **Corbis: Jose Fuste Raga (cr). Dorling Kindersley:** The British Museum (tr). **240-241 Alamy Images:** Liu Xiaoyang / China Images (c). **240 ChinaFotoPress: (tl). Dorling Kindersley:** By permission of The British Library (bl). **241 Corbis:** Nik Wheeler (br). **242 Alamy Images:** The Art Gallery Collection (tr). **Corbis:** George Steinmetz (bl). **Dorling Kindersley:** Demetrio Carrasco (c) CONACULTA-INAH-MEX; / CONACULTA-INAH-MEX. Authorized reproduction by the Instituto Nacional de Antropología e Historia (tl). **243 Corbis:** Charles & Josette Lenars.

All other images © Dorling Kindersley
For further information see:
www.dkimages.com

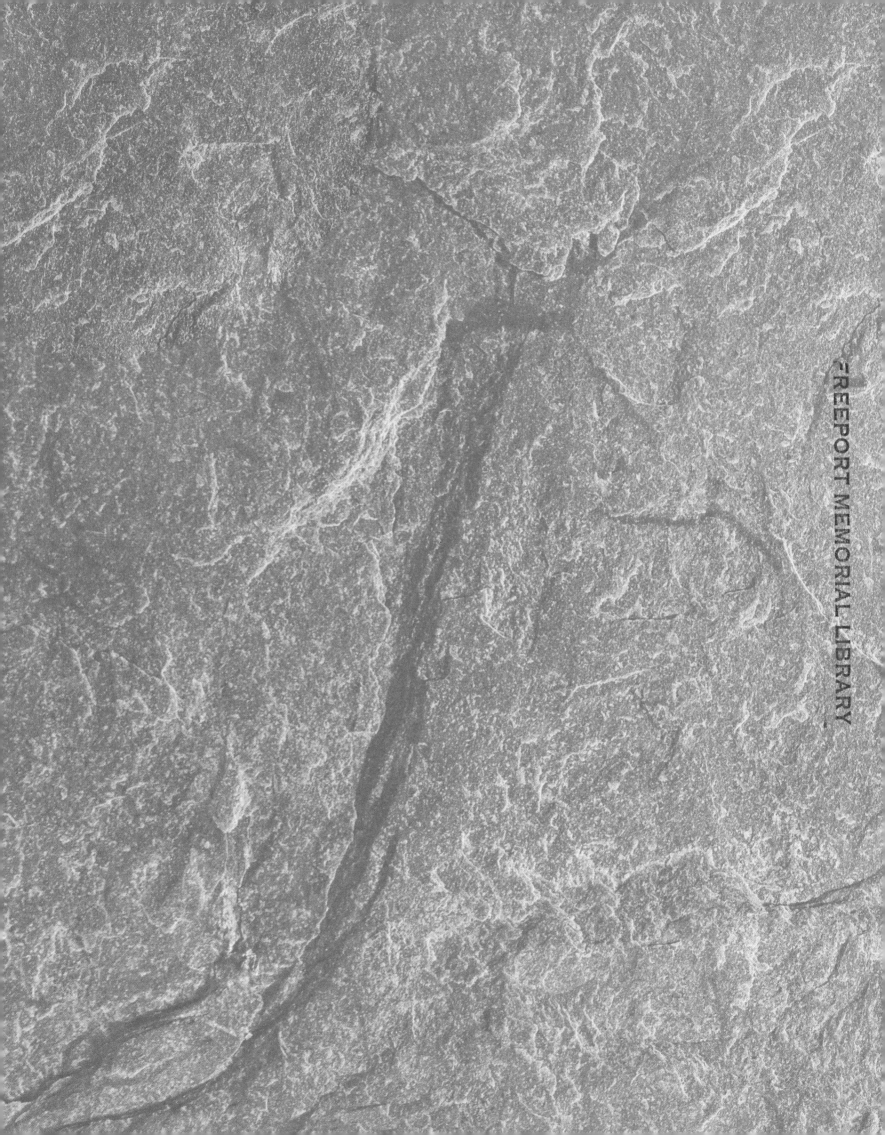